Key Methods and Concepts in Condensed Matter Physics

Green's functions and real space renormalization group

Key Methods and Concepts in Condensed Matter Physics

Green's functions and real space renormalization group

Mucio A Continentino

Centro Brasileiro de Pesquisas Físicas, Rua Dr. Xavier Sigaud, 150 - Urca, Rio de Janeiro, Brazil

IOP Publishing, Bristol, UK

ISBN 978-0-7503-3395-5 (ebook)
ISBN 978-0-7503-3393-1 (print)
ISBN 978-0-7503-3396-2 (myPrint)
ISBN 978-0-7503-3394-8 (mobi)

DOI 10.1088/978-0-7503-3395-5

Version: 20210401

IOP ebooks

British Library Cataloguing-in-Publication Data: A catalogue record for this book is available from the British Library.

Published by IOP Publishing, wholly owned by The Institute of Physics, London

IOP Publishing, Temple Circus, Temple Way, Bristol, BS1 6HG, UK

US Office: IOP Publishing, Inc., 190 North Independence Mall West, Suite 601, Philadelphia, PA 19106, USA

To Sonia, Michel, David and Alan.

Contents

Preface

About this book

This book presents an introduction to key concepts and methods in condensed matter physics and statistical mechanics. The connection between these two fields is an old, and deep one. While the latter is mostly a theoretical field, the former could not progress without the tools and ideas provided by statistical mechanics. This also provides a framework to rationalize experimental results in solid state physics and to make predictions. This text is a reaffirmation of this close and fruitful relationship. Motivated by concrete problems in condensed matter we introduce some useful theoretical frameworks that allow us to help understand these results. Two main techniques of statistical mechanics are presented, the Green's function method and the renormalization group (RG). Green's functions provide a powerful way to obtain correlation functions that in turn are directly related to experimental quantities. We introduce this technique with a pragmatical approach considering real problems and applications. The renormalization group provides the modern explanation of critical phenomena, but its concepts extend much beyond this field. We introduce it in its more intuitive version in real space. This approach is not always appropriate to describe systems in real lattices for which they provide an uncontrolled approximation. With this proviso it presents such a nice illustration of the RG and its concepts of fixed points, flows in parameter space, crossover and universality that it justifies studying it. Besides, there is a class of lattices named *hierarchical lattices* for which this real space version of the RG provides exact results. In some cases these lattices are even good approximations for real materials.

Acknowledgements

I would like to thank my colleagues, post-doctoral researchers, and students for discussions on the topics of this book. I am grateful to Conselho Nacional de Pesquisas - CNPq and Fundação de Amparo a Pesquisa do Estado do Rio de Janeiro - FAPERJ for partial financial support.

My special thanks to Emily Tapp and Robert Trevelyan of IOP Publishing for their assistance.

Author biography

Mucio A Continentino

Mucio A Continentino is a Full Researcher at Centro Brasileiro de Pesquisas Físicas in Rio de Janeiro. He is a member of the Brazilian Academy of Sciences and has authored the book *Quantum Scaling in Many-body Systems: an Approach to Quantum Phase Transitions* published by Cambridge University Press.

IOP Publishing

Key Methods and Concepts in Condensed Matter Physics
Green's functions and real space renormalization group
Mucio A Continentino

Chapter 1

The fluctuation–dissipation theorem

1.1 Introduction

In this chapter we introduce an important tool that will be used in a large part of this book. This is the method of Green's functions [1–3] that together with the fluctuation–dissipation theorem [4, 5] allow us to calculate correlation functions, which are directly related to experimental quantities.

1.2 Linear response theory

Let us imagine a situation, which resembles very much that of a real experiment. Consider a physical system that we want to study and for this purpose we couple it to a signal generator (SG), as shown in figure 1.1. The SG is going to disturb the system through its coupling to one of its physical variables, like magnetic moment or charge. We want to obtain the system's response due to this external perturbation to get information about the system and its spontaneous fluctuations in the absence of the perturbation. For this purpose we assume the perturbation is weak and we are going to develop what is known as a *linear response theory*. We can think of shooting neutrons that couple to the magnetic moments of the material or applying a voltage that will give rise to an electric field that acts in its charge degrees of freedom.

The total Hamiltonian describing the system, the signal generator and their coupling is given, respectively, by,

$$H = H_0 + H_{SG} + V.$$

The signal generator is assumed not to be affected by the excitations it produces in the system. It is essentially a classical object with a few degrees of freedom. In this case, the commutator $[H_0 + V, H_{SG}] = 0$ and we can simply write, $H = H_0 + V$. The term in the Hamiltonian that takes into account the interaction between the system and the SG is written as $V(t) = -\mathbf{A}F(t)$ where $\mathbf{A}$ is an operator representing a physical variable of the system that couples to the SG. The function $F(t)$ is a c-number and gives the time dependence of the signal. We assume this perturbation

doi:10.1088/978-0-7503-3395-5ch1

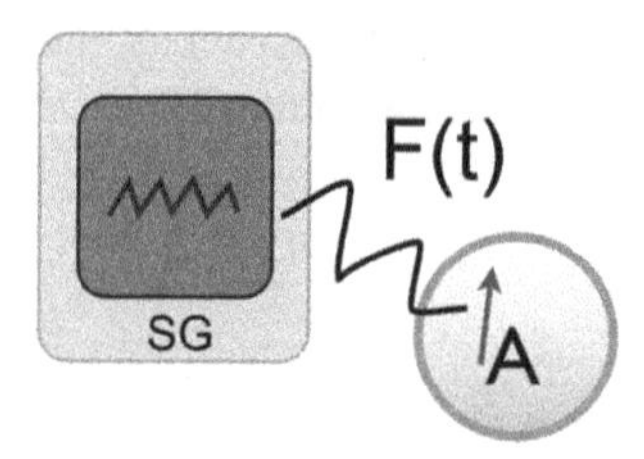

Figure 1.1. The signal generator (SG) couples to the physical variable **A** of the system causing a perturbation.

is applied adiabatically at $t = -\infty$. This is implemented multiplying $F(t)$ by a factor $e^{t/\tau}$ where τ is a positive constant and $t \leqslant 0$.

Representations

In the usual Schrödinger's representation, the states, represented by *kets*, evolve in time according to Schrödinger's equation (we use $\hbar = 1$, but in some situations we may restore it explicitly). We have,

$$i\frac{\partial}{\partial t}|\psi(t)\rangle_S = H|\psi(t)\rangle_S = (H_0 + V(t))|\psi(t)\rangle_S.$$

In this representation the dynamics of the system is contained in the time dependence of the kets. For a time independent Hamiltonian, the operators are constant in time. It is useful to introduce new representations, as the *interaction representation* [2, 3] where both, states and operators associated with physical variables change in time. In this *interaction representation* the states are written as

$$|\psi(t)\rangle_I = e^{iH_0t}|\psi(t)\rangle_S.$$

We can easily find that these states obey the following modified Schrödinger equation,

$$i\frac{\partial}{\partial t}|\psi(t)\rangle_I = V_I(t)|\psi(t)\rangle_I, \tag{1.1}$$

where

$$V_I(t) = e^{iH_0t}Ve^{-iH_0t}.$$

Notice that even if V is a time independent perturbation it now acquires a time dependence. The dynamics of an operator $B_I(t) = e^{iH_0t}Be^{-iH_0t}$ in the interaction representation is governed by the following equation of motion

$$i\frac{d}{dt}B_I(t) = [B_I(t), H_0], \tag{1.2}$$

where $[B_I, H_0] = B_IH_0 - H_0B_I$ is the commutator of these operators. In the case where B has an explicit time dependence, we have an additional term in this equation given by, $\partial B/\partial t$.

A formal solution of equation (1.1) can be obtained in terms of an evolution operator,

$$|\psi(t)\rangle_I = U(t)|\psi(t = -\infty)\rangle_I, \tag{1.3}$$

where

$$U(t) = \mathcal{T}e^{-\frac{i}{\hbar}\int_{-\infty}^{t} dt' V_I(t')} \tag{1.4}$$

and $U(-\infty) = 1$. The symbol $\mathcal{T}$ represents the *time ordering operator*. The first terms in the expansion of $U(t)$ are,

$$U(t) = 1 + \frac{-i}{\hbar}\int_{-\infty}^{t} dt' V_I(t') + \left(\frac{-i}{\hbar}\right)^2 \int_{-\infty}^{t} dt' \int_{-\infty}^{t'} dt'' V_I(t')V_I(t'') + \cdots.$$

The last term can be formally rewritten as

$$\int_{-\infty}^{t} dt' \int_{-\infty}^{t'} dt'' V_I(t')V_I(t'') = \frac{1}{2}\int_{-\infty}^{t} dt' \int_{-\infty}^{t} dt'' \mathcal{T}(V_I(t')V_I(t'')).$$

The role of $\mathcal{T}(\prod_i X_i(t_i))$, as a time ordering operator, is to write the operators inside the brackets in decreasing time order, i.e., later times to the left. For the second order term we have,

$$\mathcal{T}(V_I(t')V_I(t'')) = \theta(t' - t'')(V_I(t')V_I(t'')) + \theta(t'' - t')(V_I(t'')V_I(t')),$$

where $\theta(t)$ is the Heaviside step-function,

$$\theta(t) = \begin{cases} 1 & \text{for} \quad t \geqslant 0 \\ 0, & \text{for} \quad t < 0. \end{cases}$$

The adiabatic condition implies that the perturbation $V(t)$ only becomes important near $t \approx 0$. Consequently the evolution operator only deviates significantly from unity for $t \gtrsim 0$. For this reason, we can replace in equation (1.3), $|\psi(t = -\infty)\rangle_I \to |\psi(t = 0)\rangle_I$. Notice that

$$|\psi(t = 0)\rangle_I = |\psi(t = 0)\rangle_S \equiv |\psi\rangle_I^0, \tag{1.5}$$

where $|\psi\rangle_I^0$ is an eigenstate of the unperturbed Hamiltonian H_0.

We remark that the expectation value of an operator B in the Schrödinger representation is the same as in the interaction representation, since

$$\begin{aligned} {}_S\langle\psi(t)|B|\psi(t)\rangle_S &= {}_S\langle\psi(t)|e^{-iH_0t}e^{iH_0t}Be^{-iH_0t}e^{iH_0t}|\psi(t)\rangle_S \\ &= {}_I\langle\psi(t)|B_I(t)|\psi\rangle_I. \end{aligned} \tag{1.6}$$

1.2.1 Response of the system to an external perturbation

The signal generator disturbs the system through its coupling to the physical variable A. This is described by the Hamiltonian $V(t) = -\mathbf{A}F(t)$. Our main goal now is to calculate the change ΔB of the expectation value of a physical quantity B of the system due to the perturbation caused by the SG. Notice that the physical variable B is not necessarily the one that couples to the SG, in the present case A. We assume the perturbation is weak and calculate the change ΔB to first order perturbation theory in the strength of the signal $F(t)$. For convenience we use the interaction representation. We get to first order in F,

$$\begin{aligned} {}_I\langle\psi(t)|B_I(t)|\psi(t)\rangle_I &= \left({}^0_I\langle\psi(t)| + {}^1_I\langle\psi(t)|\right)B_I(t)\left(|\psi(t)\rangle^0_I + |\psi(t)\rangle^1_I\right) \\ &= {}^0_I\langle\psi|B_I|\psi\rangle^0_I + {}^1_I\langle\psi|B_I|\psi\rangle^0_I + {}^0_I\langle\psi|B_I|\psi\rangle^1_I + O(F^2). \end{aligned} \tag{1.7}$$

The superscript in the ket refers to the order of the perturbation and the subscript to the representation. The change of the expectation value of the physical variable B to first order in the perturbation is given by,

$$\Delta B \equiv {}_I\langle\psi(t)|B_I(t)|\psi(t)\rangle_I - {}^0_I\langle\psi(t)|B|\psi(t)\rangle^0_I,$$

which together with equation (1.7) yields,

$$\Delta B = {}^1_I\langle\psi(t)|B|\psi(t)\rangle^0_I + {}^0_I\langle\psi(t)|B|\psi(t)\rangle^1_I.$$

Using equation (1.3) and the first order term in the expansion of $U(t)$, equation (1.4), we get

$$|\psi(t)\rangle^1_I = \frac{-1}{i}\int_{-\infty}^{t} dt' A_I(t')F(t')|\psi\rangle^0_I.$$

Then,

$$\begin{aligned} \Delta B &= \frac{1}{i}\int_{-\infty}^{t} dt'\ {}^0_I|(A_I(t')B_I(t) - B_I(t)A_I(t'))|\psi\rangle^0_I F(t') \\ &= \frac{1}{i}\int_{-\infty}^{t} dt'\ {}^0_I\langle\psi|[A_I(t'), B_I(t)]|\psi\rangle^0_I F(t'). \end{aligned} \tag{1.8}$$

Using the step-function, $\theta(t)$ we can write,

$$\Delta B = \int_{-\infty}^{\infty} dt' \chi_{BA}(t - t')F(t'), \tag{1.9}$$

where we used that,

$${}^0_I\langle\psi|[A_I(t'), B_I(t)]|\psi\rangle^0_I = {}^0_I\langle\psi|[A_I(0), B_I(t - t')]|\psi\rangle^0_I \tag{1.10}$$

is a function only of $(t - t')$.

The zero temperature, time dependent susceptibility $\chi_{BA}(t - t')$ is given by,

$$\chi_{BA}(t - t') = \frac{i}{\hbar}\theta(t - t')\ {}^0_I\langle\psi|[B_I(t - t')A_I(0)]|\psi\rangle^0_I. \tag{1.11}$$

The susceptibility is a *Green's function,* which is essentially a causal expectation value of a commutator. At this order in perturbation theory, it depends only on the parameters of the unperturbed system, i.e., on the Hamiltonian H_0 and its eigenstates $|\psi(t=0)\rangle_I = |\psi(t=0)\rangle_S \equiv |\psi\rangle_I^0$. We reintroduced $\hbar$ above to remark that χ_{BA} is a genuine quantum object, the commutator being ill-defined in the classical limit.

For a perturbation $F(t) = F_0\delta(t - t_0)$, the response of the system can be obtained from equation (1.9) as,

$$\Delta B(t) = \chi_{BA}(t - t_0)F_0$$

so that χ_{BA} is indeed a susceptibility.

The convolution theorem allows us to rewrite equation (1.9) as,

$$\Delta B(\omega) = \chi_{BA}(\omega)F(\omega), \tag{1.12}$$

where

$$\chi_{BA}(\omega) = \int_{-\infty}^{\infty} dt e^{i\omega t}\chi_{BA}(t).$$

$\Delta B(\omega)$ and $F(\omega)$ are Fourier transforms, with similar definitions. Notice that if $F(t)$ is real, this implies that $\chi(t)$ is also real. Consequently we find, $\chi_{BA}(-\omega) = \chi_{BA}^*(\omega)$.

1.3 Fluctuation–dissipation theorem

1.3.1 Dissipation

When the signal generator disturbs the system it transfers energy to it. This energy absorbed by the system is eventually dissipated. The energy loss of the signal generator per unit time is given by $Q = d\Delta\bar{E}/dt$ where $\Delta\bar{E}$ is the expectation value of $\Delta H = H - H_0 = -\Delta AF(t)$. Then, we have

$$Q = \frac{\partial \Delta \mathcal{H}}{\partial t} = -\Delta A(t)\frac{\partial F}{\partial t},$$

where

$$\Delta A(t) = \int_{-\infty}^{\infty} dt' \chi_{AA}(t - t')F(t')$$

and

$$\chi_{AA}(t - t') = i\theta(t - t')\, {}_I^0\langle\psi|[A_I(t - t'), A_I(0)]|\psi\rangle_I^0.$$

Let us take $F(t) = F_0 \cos \omega t = (F_0/2)(e^{i\omega t} + e^{-i\omega t})$. In this case,

$$\Delta A = \int_{-\infty}^{\infty} dt' \chi_{AA}(t - t')\frac{F_0}{2}(e^{i\omega t'} + e^{-i\omega t'}) = \frac{F_0}{2}[\chi_{AA}(-\omega)e^{i\omega t} + \chi_{AA}(\omega)e^{-i\omega t}].$$

Then,

$$Q = \frac{F_0}{2}[\chi^*_{AA}(\omega)e^{i\omega t} + \chi_{AA}(\omega)e^{-i\omega t}]\frac{\omega F_0}{2i}(e^{i\omega t} + e^{-i\omega t}).$$

The average energy absorbed by the system in a cycle of period $T = 2\pi/\omega$, $\bar{Q} = \int_0^T Q(t)dt$ is given by,

$$\bar{Q} = -\frac{F_0^2\omega}{4i}[\chi^*_{AA}(\omega) - \chi_{AA}(\omega)]$$

or

$$\bar{Q} = \frac{F_0^2}{2}\omega \,\mathrm{Im}\,\chi_{AA}(\omega). \tag{1.13}$$

Since $\bar{Q} > 0$, $\omega \,\mathrm{Im}\,\chi_{AA}(\omega) > 0$, implying that $\mathrm{Im}\,\chi_{AA}(\omega)$ is an odd function of ω. The energy absorbed from the SG is dissipated by the system and is given in terms of the imaginary part of a dynamical susceptibility.

1.3.2 Fluctuations

Now we turn off the signal generator ($V = 0$) and concentrate on the *spontaneous fluctuations* of the system. For a given variable A these fluctuations can be characterized by a quantity like $\langle(A^2 - \langle A\rangle^2)\rangle$, where the symbol $\langle\cdots\rangle$ stands for $\langle\psi|\cdots|\psi\rangle$, where $|\psi\rangle = |\psi\rangle_I^0$ is an eigenstate of H_0. We will consider a more general *correlation function* between two physical variables of the system defined by,

$$K_{BA}(t) = \frac{1}{2}\langle\psi|(B(t) - \langle B\rangle)(A(0) - \langle A\rangle) + (A(0) - \langle A\rangle)(B(t) - \langle B\rangle)|\psi\rangle.$$

For simplicity we assume that the expectation values of A and B vanish, i.e., $\langle A\rangle = \langle B\rangle = 0$ and write

$$K_{BA}(t) = \frac{1}{2}\langle\psi|\{B(t), A(0)\}|\psi\rangle, \tag{1.14}$$

where $\{B, A\} = BA + AB$ is the *anticommutator* of the operators B and A.

The fluctuation–dissipation (FD) theorem [4, 5], establishes a relation between the spontaneous fluctuations of the system as characterized by equation (1.14) and a susceptibility,

$$\chi_{BA}(t) = (1/2)\theta(t)\langle\psi|[B(t), A(0)]|\psi\rangle, \tag{1.15}$$

which gives the response of the system to a weak external stimulus due to the signal generator. We used above that K_{BA} and χ_{BA} are functions of $(t - t')$ and took $t' = 0$. From a mathematical point of view, the FD theorem provides a relationship between the expectation value of an anticommutator $K_{BA}(t)$ and that of a causal commutator $\chi_{BA}(t)$.

Next, we will derive the FD theorem, but instead of working at zero temperature ($T = 0$) as we have done so far, we will consider its generalization to finite T. In this case the expectation values are replaced by averages over a canonical or grand-canonical ensembles, i.e.,

$$\langle\psi|C|\psi\rangle \rightarrow \frac{Tre^{-\beta H_0}C}{Tre^{-\beta H_0}}.$$

Using the eigenstates of H_0, we can write the trace (Tr) as,

$$\frac{Tre^{-\beta H_0}C}{Tre^{-\beta H_0}} = \frac{\sum_n \langle n|e^{-\beta H_0}C|n\rangle}{\sum_n \langle n|e^{-\beta H_0}|n\rangle} = \frac{\sum_n \langle n|C|n\rangle e^{-\beta E_n}}{\sum_n e^{-\beta E_n}}.$$

Let us consider first the correlation function of the spontaneous fluctuations,

$$K_{BA}(t) = \frac{Tre^{-\beta H_0}\frac{1}{2}\{B(t), A(0)\}}{Tre^{-\beta H_0}} = \frac{\sum_n e^{-\beta E_n}\langle n|B(t)A + AB(t)|n\rangle}{\sum_n e^{-\beta E_n}}.$$

Inserting a complete set of eigenstates, $\sum_m |m\rangle\langle m| = 1$ between the operators and recalling that $B(t) = e^{iH_0 t}Be^{-iH_0 t}$ and that $H_0|m\rangle = E_m|m\rangle$, we get,

$$K_{BA}(t) = \frac{1/2}{Z_0}\sum_{m,n} e^{-\beta E_n}[e^{i\omega_{nm}t}B_{nm}A_{mn} + e^{-i\omega_{nm}t}A_{nm}B_{mn}], \tag{1.16}$$

where $Z_0 = \sum_n e^{-\beta E_n}$ is the partition function of the system, $B_{nm} = \langle n|B|m\rangle$ and $\omega_{nm} = E_n - E_m$. Introducing $K_{BA}(\omega) = \int_{-\infty}^{\infty} K_{BA}(t)e^{i\omega t}dt$, we obtain the *spectral decomposition* of the correlation function,

$$K_{BA}(\omega) = \frac{1/2}{Z_0}\sum_{m,n} e^{-\beta E_n} \times [B_{nm}A_{mn}\delta(\omega + \omega_{nm}) + A_{nm}B_{mn}\delta(\omega - \omega_{nm})] \tag{1.17}$$

We also define the *spectral density* of a product of operators by,

$$I_{BA}(\omega) = \frac{1}{Z_0}\sum_{m,n} e^{-\beta E_n}[B_{nm}A_{mn}\delta(\omega + \omega_{nm})]. \tag{1.18}$$

We now obtain the spectral decomposition of the response function $\chi_{BA}(\omega)$. Since

$$\chi_{BA}(t) = i\theta(t)\frac{Tre^{-\beta H_0}[B(t)A(0)]}{Z_0} \tag{1.19}$$

and proceeding as we did before for the correlation function, we get,

$$\chi_{BA}(t) = \frac{i}{\hbar}\frac{\theta(t)}{Z_0}\sum_{m,n} e^{-\beta E_n}[e^{i\omega_{nm}t}B_{nm}A_{mn} - e^{-i\omega_{nm}t}A_{nm}B_{mn}], \tag{1.20}$$

where we reintroduced $\hbar$ for completeness. The main difference with the previous calculation for the correlation function appears when we take the time Fourier transform to obtain $\chi_{BA}(\omega)$. Now we have to deal with the following integral,

$$\int_{-\infty}^{\infty} dt\theta(t)e^{i(\omega+\omega_{nm})t}.$$

For this purpose we write the following formal representation for the step function,

$$\theta(t) = \begin{cases} e^{-\epsilon t}(\epsilon \to 0, \epsilon > 0), & \text{for} \quad t > 0 \\ 0, & \text{for} \quad t < 0. \end{cases}$$

Substituting, we get,

$$\int_{-\infty}^{\infty} dt\theta(t)e^{i(\omega+\omega_{nm})t} = \lim_{\epsilon\to 0}\int_{0}^{\infty} dte^{i(\omega+\omega_{nm}+i\epsilon)t}.$$

The factor $\epsilon > 0$ guarantees the convergence of the integral at infinity and this can be easily calculated. We find,

$$\int_{-\infty}^{\infty} dt\theta(t)e^{i(\omega+\omega_{nm})t} = \lim_{\epsilon\to 0}\frac{-1}{i(\omega + \omega_{nm} + i\epsilon)}.$$

The response function can then be written as,

$$\chi_{BA}(\omega) = \frac{-1}{Z_0}\sum_{m,n} e^{-\beta E_n}\left[\frac{B_{nm}A_{mn}}{\omega + \omega_{nm} + i\epsilon} - \frac{A_{nm}B_{mn}}{\omega - \omega_{nm} + i\epsilon}\right].$$

Using the Dirac relation,

$$\lim_{\epsilon\to 0}\frac{1}{(\omega + \omega_{nm} \pm i\epsilon)} = P\frac{1}{\omega + \omega_{nm}} \mp i\pi\delta(\omega + \omega_{nm}), \tag{1.21}$$

where P is the principal part, we can obtain the imaginary part of $\chi_{BA}(\omega)$ that as we have seen yields the energy dissipated by the system,

$$\mathrm{Im}\,\chi_{BA}(\omega) = \frac{\pi}{\hbar}\frac{1}{Z_0}\sum_{m,n} e^{-\beta E_n} \times [B_{nm}A_{mn}\delta(\omega + \omega_{nm}) - A_{nm}B_{mn}\delta(\omega - \omega_{nm})].$$

Let us summarize the main results obtained so far,

$$K_{BA}(\omega) = \frac{1}{2}\sum_{m,n}\rho_n[B_{nm}A_{mn}\delta(\omega + \omega_{nm}) + A_{nm}B_{mn}\delta(\omega - \omega_{nm})]$$

$$\mathrm{Im}\,\chi_{BA}(\omega) = \frac{\pi}{\hbar}\sum_{m,n}\rho_n[B_{nm}A_{mn}\delta(\omega + \omega_{nm}) - A_{nm}B_{mn}\delta(\omega - \omega_{nm})],$$

where $\rho_n = e^{-\beta E_n}/Z_0$. However, $\rho_n\delta(\omega + \omega_{nm}) = e^{\beta\omega}\rho_m\delta(\omega + \omega_{nm})$ and dealing carefully with indices we obtain,

$$K_{BA}(\omega) = \frac{1}{2}(1 + e^{\beta\omega})\sum_{m,n}\rho_n B_{nm}A_{mn}\delta(\omega + \omega_{nm})$$

$$\mathrm{Im}\,\chi_{BA}(\omega) = \frac{\pi}{\hbar}(e^{\beta\omega} - 1)\sum_{m,n}\rho_n B_{nm}A_{mn}\delta(\omega + \omega_{nm}),$$

where we notice the appearance of the spectral density $I_{BA}(\omega)$, given by equation (1.18). Relating the values of the spectral density in the equations above we obtain,

$$\frac{2K_{BA}(\omega)}{1 + e^{\beta\omega}} = \frac{\hbar}{\pi}\frac{\mathrm{Im}\,\chi_{BA}(\omega)}{e^{\beta\omega} - 1}$$

or the most common form of the FD theorem:

$$K_{BA}(\omega) = \frac{\hbar}{2\pi}\coth\left(\frac{\beta\hbar\omega}{2}\right)\mathrm{Im}\,\chi_{BA}(\omega). \tag{1.22}$$

In the classical limit, $(\hbar\omega/k_BT) \ll 1$, we get,

$$K_{BA}(\omega) = \frac{k_BT}{\pi}\frac{\mathrm{Im}\,\chi_{BA}(\omega)}{\omega}.$$

Notice that $\hbar$ cancels out as expected. In the limit $T \to 0$, we find,

$$K_{BA}(\omega) = \frac{\hbar}{2\pi}\mathrm{Im}\,\chi_{BA}(\omega)$$

that refers to the zero point fluctuations of the system.

It is also interesting to consider the time dependence of the fluctuations. If we write

$$K_{BA}(t) = \int \Gamma(t - t')\chi''_{BA}(t')dt', \tag{1.23}$$

where $\chi''_{BA}(t') = \int d\omega\, \exp(i\omega t) Im\chi_{BA}(\omega)$, we get

$$\Gamma(t) = \frac{k_BT}{2\pi}\coth\left(\frac{\pi k_BT}{\hbar}t\right).$$

In the classical limit, $\hbar \to 0$,

$$\Gamma(t) = \frac{k_B T}{\sqrt{2\pi}}(1 + 2e^{-\frac{t}{\tau}}),$$

where the characteristic time for the decay of spontaneous fluctuations

$$\tau = \frac{\hbar}{2\pi k_B T}.$$

On the other hand in the quantum limit,

$$\Gamma(t) = \frac{\hbar}{\sqrt{2}\pi^{3/2}}\frac{1}{t},$$

and the decay of the fluctuations is very slow, varying algebraically with time.

1.3.3 Analytic properties of the Green's function

Consider the function in the complex z-plane,

$$\chi(z) = \int_{-\infty}^{\infty} dt e^{izt} \chi(t)$$

with

$$\chi(t) = 0 \text{ for } \quad t < 0. \tag{1.24}$$

Lemma 1. *$\chi(z)$ is analytic in the upper part of the complex plane.*

Proof 1. Let us define, $z = \omega + i\epsilon$, we have

$$\chi(z) = \int_{-\infty}^{\infty} dt e^{i\omega t} e^{-\epsilon t} \chi(t).$$

The factor $e^{-\epsilon t}$ guarantees the convergence of the integral at $t = +\infty$. On the other hand the step function $\theta(t)$ (see equation (1.24)) guarantees that there is no problem of convergence at $t = -\infty$. Then $\chi(z)$ is analytical for $z = \omega + i\epsilon$, i.e., in the upper part of the complex plane. □

Kramers–Kronig relations

Consider the integral below along the closed contour shown in figure 1.2,

$$I = \oint dz \frac{\chi(z)}{\omega - z - i\epsilon}.$$

The integrand has a pole at $z = \omega - i\epsilon$. Since $\chi(z)$ is analytical in the upper half-plane, the integral above by Cauchy's theorem vanishes. Also, because $\chi(|z|) \to 0$ when $|z| \to \infty$, the integral along the arc vanishes. In this case we get,

$$I = \int_{-\infty}^{\infty} d\omega' \frac{\chi(\omega')}{\omega - \omega' - i\epsilon} = 0.$$

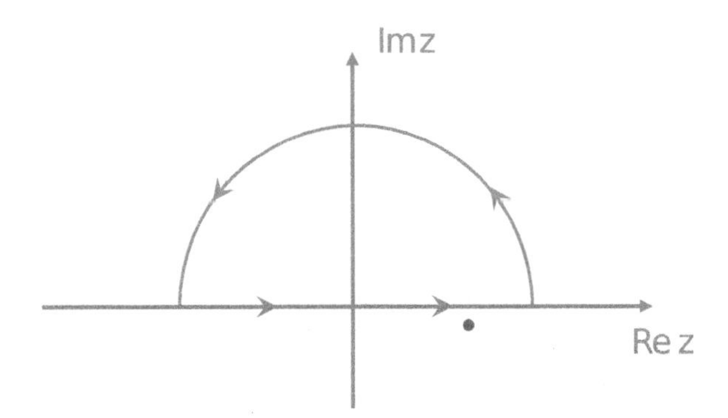

Figure 1.2. The contour of integration in the complex plane and the pole (red dot).

We now write $\chi(\omega) = \mathrm{Re}\,\chi(\omega) + i\ \mathrm{Im}\,\chi(\omega)$ and using that,

$$\frac{1}{\omega - \omega' \pm i\epsilon} = P\frac{1}{\omega - \omega'} \mp i\pi\delta(\omega - \omega'),$$

we have

$$\int_{-\infty}^{\infty} d\omega'[\mathrm{Re}\,\chi(\omega') + i\ \mathrm{Im}\,\chi(\omega')][\mathrm{P}\frac{1}{\omega - \omega'} + i\pi\delta(\omega - \omega')] = 0.$$

Since both real and imaginary parts must be zero, we get the *Kramers–Kronig relations*,

$$\mathrm{Im}\,\chi(\omega) = \frac{1}{\pi}\mathrm{P}\int_{-\infty}^{\infty} \mathrm{d}\omega' \frac{\mathrm{Re}\,\chi(\omega')}{\omega - \omega'} \tag{1.25}$$

$$\mathrm{Re}\,\chi(\omega) = -\mathrm{P}\frac{1}{\pi}\int_{-\infty}^{\infty} \mathrm{d}\omega' \frac{\mathrm{Im}\,\chi(\omega')}{\omega - \omega'}. \tag{1.26}$$

References

[1] Tyablikov S V 1967 *Methods in the Quantum Theory of Magnetism* (New York: Plenum Press)
[2] Fetter A L and Walecka J D 2003 *Quantum Theory of Many-Particle Systems* (Dover Books on Physics) (New York: Dover Publications Inc.)
[3] Kadanoff L P and Baym G 1962 *Quantum Statistical Mechanics, Green's Function Methods in Equilibrium and Nonequilibrium Problems* (New York: W. A. Benjamin, Inc.)
[4] Callen H B and Welton T A 1951 *Phys. Rev.* **83** 34
[5] Callen H B 1962 *Fluctuation, Relaxation and Resonance in Magnetic Systems* ed D Ter Haar (Edinburgh: Oliver and Boyd)

IOP Publishing

Key Methods and Concepts in Condensed Matter Physics
Green's functions and real space renormalization group
Mucio A Continentino

Chapter 2

Green's function method

2.1 Green's functions and correlations

The physical quantities measured in experiments are generally associated with correlations functions. The Green's function method is important because it provides a very useful way to obtain these quantities. It allows the use of Feynman diagrams and it is easier to obtain the Green's function and from them the correlation functions than calculating the latter directly. In this section we introduce the Green's function method and show how to obtain from it the correlation functions [1].

Consider initially the expression for an operator in the *Heisenberg representation*,

$$A(t) = e^{iHt}Ae^{-iHt},$$

where $H = H_0 - \mu N$, with μ the chemical potential and N the total number of particles. The equation of motion for this operator is given by,

$$i\frac{dA}{dt} = [A(t), H]. \qquad (2.1)$$

Now in the grand-canonical ensemble the thermodynamic average of the physical quantity A is given by,

$$\langle A \rangle = \frac{Tre^{-\beta H}A}{Tre^{-\beta H}} = Tr\rho A,$$

where $\rho = e^{-\beta H}/Tre^{-\beta H}$ is the density matrix and Tr means the trace operation.

We define two types of Green's functions namely, retarded and advanced, given respectively by [1],

$$G^r_{AB}(t, t') = \langle\langle A(t); B(t')\rangle\rangle^r = \theta(t - t')\langle[A(t), B(t')]_\eta\rangle$$
$$G^a_{AB}(t, t') = \langle\langle A(t); B(t')\rangle\rangle^a = -\theta(t' - t)\langle[A(t), B(t')]_\eta\rangle.$$

The symbol, $[A, B]_\eta = AB - \eta BA$ where $\eta = -1$ for fermionic operators and $\eta = +1$ for bosons.

It is also useful to define a causal Green's function by,

$$G^c_{AB}(t, t') = \langle\langle A(t); B(t')\rangle\rangle^c = \langle T_\eta A(t)B(t')\rangle,$$

where $T_\eta A(t)B(t') = \theta(t - t')A(t)B(t') + \eta\theta(t' - t)B(t')A(t)$ is the time ordering operator, in this case acting on $A(t)B(t')$. This type of Green's function or propagator is particularly useful for perturbations expansions as they can be represented in terms of Feynman diagrams.

Since the trace operation is invariant to cyclic permutations, it is straightforward to show that for time independent Hamiltonians H, the Green's function are time translation invariant and depend only on $(t - t')$.

Using equation (2.1) it is easy to show that both, retarded and advanced Green's function obey the following equation of motion.

$$i\frac{d}{dt}\langle\langle A(t); B(t')\rangle\rangle = i\delta(t - t')\langle [A(t), B(t')]_\eta\rangle + \langle\langle [A(t), H]; B(t')\rangle\rangle,$$

where we used the representation of the step function, $\theta(t) = \int_{-\infty}^{t} e^{\epsilon t'}\delta(t')dt'$ ($\epsilon \to 0^+$) such that $d\theta(t)/dt = \delta(t)$.

In general the term $\langle\langle [A(t), H]; B(t')\rangle\rangle$ gives rise to new Green's functions that are of higher order in the number of operators. This is the case for the interacting many-body problems that we are interested in. Then, if we want to calculate $\langle\langle A(t); B(t')\rangle\rangle$, we must write the equations of motion for the new Green's functions and this procedure generates an infinite chain of coupled equations that needs to be solved. This chain can be interrupted if at some level we *decouple* the new Green's function into a lower order one, as will be shown further on.

The time Fourier transform of the Green's functions are given by,

$$G_{AB}(\omega) = \langle\langle A; B\rangle\rangle_\omega = \frac{1}{2\pi}\int_{-\infty}^{+\infty} dt G_{AB}(t)e^{i\omega t},$$

where we used time translation invariance. In terms of these, the equation of motion is given by,

$$\omega\langle\langle A; B\rangle\rangle_\omega = \frac{i}{2\pi}\langle [A, B]_\eta\rangle + \langle\langle [A, H]; B\rangle\rangle_\omega.$$

2.1.1 An important result

The spectral density of the product of operators $I_{BA}(\omega)$ defined previously in equation (1.18) allows us to write the correlation functions as,

$$\langle B(t)A(t')\rangle = \int_{-\infty}^{+\infty} d\omega I_{BA}(\omega)e^{-i\omega(t-t')}$$

$$\langle A(t)B(t')\rangle = \int_{-\infty}^{+\infty} d\omega I_{BA}(\omega)e^{\beta\omega}e^{-i\omega(t-t')},$$

such that

$$\langle [A(t), B(t')]_\eta \rangle = \int_{-\infty}^{+\infty} d\omega I_{BA}(\omega)(e^{\beta\omega} - \eta)e^{-i\omega(t-t')},$$

and for the retarded Green's function,

$$G^r_{AB}(\omega) = \frac{1}{2\pi}\int_{-\infty}^{+\infty} d\omega' I_{BA}(\omega')(e^{\beta\omega'} - \eta)\int_{-\infty}^{\infty} dt e^{i(\omega-\omega')t}\theta(t).$$

The time integral is obtained as before and we finally get the spectral representation of the retarded and advanced Green's function,

$$G^r_{AB}(\omega) = \frac{1}{2\pi}\int_{-\infty}^{+\infty} \frac{d\omega'}{\omega - \omega' + i\epsilon} I_{BA}(\omega')(e^{\beta\omega'} - \eta)$$
$$G^a_{AB}(\omega) = \frac{1}{2\pi}\int_{-\infty}^{+\infty} \frac{d\omega'}{\omega - \omega' - i\epsilon} I_{BA}(\omega')(e^{\beta\omega'} - \eta).$$

It is clear from these equations that it is convenient to abandon the indices r and a, and make an analytic continuation of $G_{AB}(\omega)$ into the complex plane, such that $G^r_{AB}(\omega) = G_{AB}(\omega + i\epsilon)$ and $G^a_{AB}(\omega) = G_{AB}(\omega - i\epsilon)$.

Notice that

$$\begin{aligned} G_{AB}(\omega + i\epsilon) - G_{AB}(\omega - i\epsilon) &= \frac{1}{2\pi}\int_{-\infty}^{+\infty} d\omega' I_{BA}(\omega')(e^{\beta\omega'} - \eta) \\ &\times \left[\frac{1}{\omega - \omega' + i\epsilon} - \frac{1}{\omega - \omega' - i\epsilon}\right] \\ &= I_{BA}(\omega)(e^{\beta\omega} - \eta) \end{aligned} \tag{2.2}$$

or,

$$I_{BA}(\omega) = \frac{1}{(e^{\beta\omega} - \eta)}[G_{AB}(\omega + i\epsilon) - G_{AB}(\omega - i\epsilon)].$$

We used the Dirac relation, equation (1.21), and remark that the principal parts cancel out in equation (2.2). Finally, using equation (2.2), we obtain a very useful result that allows us to obtain a correlation function from an appropriate Green's function,

$$\begin{aligned} \langle B(0)A(0)\rangle &= \int_{-\infty}^{+\infty} d\omega I_{BA}(\omega) \\ &= \int_{-\infty}^{+\infty} \frac{d\omega}{e^{\beta\omega} - \eta}[G_{AB}(\omega + i\epsilon) - G_{AB}(\omega - i\epsilon)]. \end{aligned} \tag{2.3}$$

This is essentially another version of the FD theorem known colloquially as the *leap theorem*. Since we will use this result frequently, we define the quantity

$$\mathcal{F}_\omega[G(\omega)] = i\lim_{\epsilon\to 0}\int_{-\infty}^{\infty}\frac{1}{e^{\beta\omega}-\eta}[G(\omega+i\epsilon)-G(\omega-i\epsilon)] \tag{2.4}$$

with $\eta = \pm 1$, for bosons and fermions, respectively.

2.1.2 Some trivial and not so trivial examples

Let us consider a gas of non-interacting spinless fermions or bosons in a lattice The Hamiltonian in both cases is given by,

$$H = \sum_{i,j} t_{ij} c_i^+ c_j,$$

where the creation c_i^+ and annihilation operators c_j, commute or anti-commute depending on whether the particles are bosons or fermions, respectively. The t_{ij} are nearest neighbors hopping terms between the sites in the lattice. They arise due to the overlap of the wave-functions in these sites. The solution to this problem corresponds to finding the average number of particles in a given site or quantum state of the lattice, $\langle n_i\rangle = \langle c_i^+ c_i\rangle$ or $\langle n_k\rangle = \langle c_k^+ c_k\rangle$, respectively. Due to translation invariance $\langle n_i\rangle = \langle n\rangle$ is site independent and the momentum k is a good quantum number. The latter correlation function can be obtained from the Green's function $\langle\langle c_k; c_k^+\rangle\rangle_\omega$ using the leap theorem.

Let us start obtaining the Green's function $\langle\langle c_i; c_j^+\rangle\rangle_\omega$ from its equation of motion,

$$\begin{aligned}\omega\left\langle\left\langle c_i; c_j^+\right\rangle\right\rangle_\omega &= \frac{1}{2\pi}\left\langle\left[c_i, c_j^+\right]_\eta\right\rangle + \left\langle\left\langle [c_i, H]; c_j^+\right\rangle\right\rangle_\omega \\ &= \frac{1}{2\pi}\delta_{ij} + \sum_l t_{il}\left\langle\left\langle c_l; c_j^+\right\rangle\right\rangle_\omega,\end{aligned} \tag{2.5}$$

where $[c_i, c_j^+]_\eta = c_i c_j^+ - \eta c_j^+ c_i = \delta_{ij}$ with $\eta = \pm 1$ for bosons and fermions, respectively. We used above the property of commutators, $[A, BC] = B[A, C] + [A, B]C$ and the commutations relations $[c_i, c_j^+] = \delta_{ij}$ for bosons and $\{c_i, c_j^+\} = \delta_{ij}$ for fermions. The symbols $[A, B] = [A, B]_+$ and $\{A, B\} = [A, B]_-$ represent commutators and anticommutators, respectively.

Since the system is non-interacting, the new generated Green's function in equation (2.5) is of the same order in second quantization operators as the original one. In k-space, as we will see below in a moment they are the same Green's function.

This equation can be solved Fourier transforming both sides in lattice space. We multiply both sides by $\sum_i e^{ik\cdot r_i}$ and $\sum_j e^{-ik'\cdot r_j}$, to obtain,

$$\omega\left\langle\left\langle c_k; c_{k'}^+\right\rangle\right\rangle_\omega = \frac{1}{2\pi}\delta_{kk'} + \epsilon_k\langle\langle c_k; c_{k'}^+\rangle\rangle_\omega. \tag{2.6}$$

Notice that

$$\sum_{i,j,l} e^{ik\cdot r_i} e^{-ik'\cdot r_j} t_{il} \left\langle \left\langle c_l;\, c_j^{+} \right\rangle \right\rangle_{\omega} = \sum_{i,l} e^{ik\cdot r_i} t_{il} \langle\langle c_l;\, c_{k'}^{+}\rangle\rangle_{\omega}$$

$$= \sum_{i,l} e^{ik\cdot(r_i - r_l)} t_{il} e^{ik\cdot r_l} \langle\langle c_l;\, c_{k'}^{+}\rangle\rangle_{\omega},$$

where $c_k = \sum_l e^{ik\cdot r_l} c_l$ and

$$\epsilon_k = \sum_i t_{li} e^{ik\cdot(r_i - r_l)}.$$

In this last sum, we used that $t_{il} = t_{li}$ and that it is independent of the particular site l due to translation invariance. For a square lattice, for example, it yields, $\epsilon_k = 2t(\cos k_x a + \cos k_y a)$ where the hopping term $t_{il} = t$ is assumed to be only between nearest neighbors and the same for both directions. For an electron gas, we generally consider $\epsilon_k = \hbar^2 k^2 / 2m$. The solution of the equation of motion, equation (2.6), can now be obtained. It is given by,

$$\langle\langle c_k;\, c_{k'}^{+}\rangle\rangle_{\omega} = \frac{1}{2\pi} \frac{\delta_{kk'}}{\omega - \epsilon_k}.$$

Notice that the poles of the Green's functions, $\omega = \epsilon_k$ correspond to the energy of the excitations of the system. This is a general result that reinforces the utility of these functions.

Finally, using the leap theorem, we get

$$\langle n_k \rangle = \frac{i}{2\pi} \int_{-\infty}^{+\infty} \frac{d\omega}{e^{\beta\omega} - \eta} \left[\langle\langle c_k;\, c_{k'}^{+}\rangle\rangle_{\omega + i\epsilon} - \langle\langle c_k;\, c_{k'}^{+}\rangle\rangle_{\omega - i\epsilon} \right] \tag{2.7}$$

or

$$\langle n_k \rangle = \frac{1}{e^{\beta\epsilon_k} - \eta},$$

where $\eta = 1$ for bosons e $\eta = -1$ for fermions. Notice that when applying the leap theorem in equation (2.7) the *principal parts of the integrals cancel out*. The occupation numbers are given by the respective Bose–Einstein and Fermi–Dirac statistical distributions. These arise exclusively from the commutation properties of the second quantization operators for fermions and bosons, respectively, i.e., $\{c_k, c_{k'}^{+}\} = \delta_{kk'}$ and $[c_k, c_{k'}^{+}] = \delta_{kk'}$.

2.1.3 Hubbard model

Mean-field approximation

The Hubbard model [2] was introduced to explain the ferromagnetism of the transition metals. Differently from the rare-earths, in these transition metals the d-electrons, which are responsible for the magnetism are itinerant due to the overlap of their wave functions in different sites. In the rare-earths, with few exceptions, the *f*-electrons are localized in the atomic sites and so are their magnetic moments.

The Hubbard Hamiltonian considers the nearest neighbors hopping t_{ij} of the *d*-electrons in the lattice and the simplest type of Coulomb correlations U among them, namely, when they occupy the same lattice site. Due to the Pauli principle these electrons must have necessarily opposite spins. The Hubbard Hamiltonian is given by,

$$H = \sum_{i,j,\sigma} t_{ij} c_{i\sigma}^{+} c_{j\sigma} + U \sum_{i} n_{i\uparrow} n_{i\downarrow}, \tag{2.8}$$

where the number of electrons at site i with spin σ is given by $n_{i\sigma} = c_{i\sigma}^{+} c_{i\sigma}$. Let us look for a ferromagnetic solution of this model, such that the correlation function $\langle c_{i\downarrow}^{+} c_{i\downarrow} \rangle$ is different from $\langle c_{i\uparrow}^{+} c_{i\uparrow} \rangle$ implying a net magnetization $m = \langle n_{i\uparrow} \rangle - \langle n_{i\downarrow} \rangle$ in the system. The correlation functions $n_{i\sigma} = \langle n_{\sigma} \rangle$ are site independent due to translation invariance and can be obtained from the Green's functions, $\langle\langle c_{i\sigma}; c_{j\sigma}^{+} \rangle\rangle_{\omega}$, which obey the following equation of motion,

$$\omega \left\langle \left\langle c_{i\sigma}; c_{j\sigma}^{+} \right\rangle \right\rangle_{\omega} = \delta_{ij} + \sum_{l} t_{il} \left\langle \left\langle c_{l\sigma}; c_{j\sigma}^{+} \right\rangle \right\rangle_{\omega} + U \left\langle \left\langle n_{i-\sigma} c_{i\sigma}; c_{j\sigma}^{+} \right\rangle \right\rangle_{\omega}. \tag{2.9}$$

The many-body term has generated a new Green's function of higher order that needs to be calculated. Another possibility is to introduce a *decoupling* in this new Green's function already at this level and write it in terms of the lower order one. In this case we write,

$$\left\langle \left\langle n_{i-\sigma} c_{i\sigma}; c_{j\sigma}^{+} \right\rangle \right\rangle_{\omega} \approx \langle n_{i-\sigma} \rangle \left\langle \left\langle c_{i\sigma}; c_{j\sigma}^{+} \right\rangle \right\rangle_{\omega} \approx \langle n_{-\sigma} \rangle \left\langle \left\langle c_{i\sigma}; c_{j\sigma}^{+} \right\rangle \right\rangle_{\omega}, \tag{2.10}$$

where in the last step we used lattice translation invariance. This decoupling corresponds to a mean field or Hartree–Fock approximation in which an electron of spin up moves in the average field produced by those of spin down and vice-versa. Finally, using Fourier transformation in lattice space, we obtain a solution for the equation of motion as,

$$\langle\langle c_{k\sigma}; c_{k'\sigma}^{+} \rangle\rangle_{\omega} = \frac{\delta_{k,k'}}{\omega - \varepsilon_k - U\langle n_{-\sigma} \rangle}. \tag{2.11}$$

Using the leap theorem, we can write,

$$\langle n_{\sigma} \rangle = \sum_{k} \langle n_{k\sigma} \rangle = \sum_{k} f(E_{k\sigma}), \tag{2.12}$$

where $f(\omega)$ is the Fermi distribution and $E_{k\sigma} = \varepsilon_k + U\langle n_{-\sigma} \rangle$ with ε_k the dispersion relation for the electrons in the lattice. Using the identity $\sum_k g(\varepsilon_k) = \sum_k \int d\omega \delta(\omega - \varepsilon_k) g(\omega)$, with the density of states $\rho(\omega) = \sum_k \delta(\omega - \varepsilon_k)$, we obtain for finite temperatures the system of equations,

$$\langle n_\uparrow \rangle = \int d\omega \frac{\rho(\omega)}{e^{\beta(\omega + U\langle n_\downarrow \rangle - \mu)} + 1}$$
$$\langle n_\downarrow \rangle = \int d\omega \frac{\rho(\omega)}{e^{\beta(\omega + U\langle n_\uparrow \rangle - \mu)} + 1}, \tag{2.13}$$

to be solved self-consistently. There is an additional condition that determines the chemical potential μ given by, $\langle n_\uparrow \rangle + \langle n_\downarrow \rangle = n$, where n is the total number of electrons per site.

As we will show next, the condition for the appearance of a ferromagnetic solution at zero temperature is given by the *Stoner criterion*,

$$U\rho(\mu) > 1$$

where $\rho(\mu)$ is the density of states of the paramagnetic metal at the Fermi level.

Nature of the mean-field ferromagnetic solution

Consider a non-interacting metallic system in the paramagnetic state. The bands of up and down spins are equally occupied up to the Fermi level ϵ_{F0} as shown in figure 2.1. For a weakly magnetized metal [3], in the presence of on-site repulsive interactions between electrons as in Hamiltonian (2.8), a small fraction of electrons is removed from the spin down band and transferred to the spin up band, as shown in figure 2.1.

Since,

$$n = \langle n_\uparrow \rangle + \langle n_\downarrow \rangle$$
$$m = \langle n_\uparrow \rangle - \langle n_\downarrow \rangle$$

the interaction term of Hamiltonian (2.8) can be written as

$$H_{\text{int}} = \frac{U}{4}(n^2 - m^2)$$

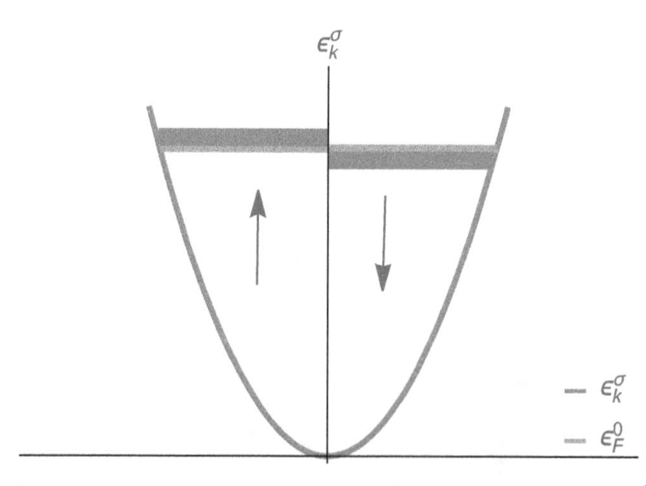

Figure 2.1. Dispersion relations for a weakly magnetized metallic system with $\langle n_\uparrow \rangle > \langle n_\downarrow \rangle$. ϵ_F^0 is the Fermi level of the paramagnetic metal. The horizontal axis is the density of states.

and the kinetic part as $H_{\text{kin}} = \sum_\sigma H_{\text{kin}}^\sigma$ where

$$H_{\text{kin}}^\sigma = \int_0^{\epsilon_F^\sigma} \rho(\epsilon)\epsilon d\epsilon.$$

Let us consider for simplicity a rectangular band with a constant density of states ρ. We have $\langle n_\sigma \rangle = \rho\epsilon_F^\sigma$ that yields

$$\epsilon_F^\uparrow + \epsilon_F^\downarrow = n/\rho$$
$$\epsilon_F^\uparrow - \epsilon_F^\downarrow = m/\rho$$

that in turn gives

$$\epsilon_F^\uparrow = \frac{n+m}{2\rho}$$
$$\epsilon_F^\downarrow = \frac{n-m}{2\rho}.$$

The total kinetic energy is given by,

$$E_{\text{kin}} = \rho\left(\frac{\epsilon_F^{\uparrow 2} + \epsilon_F^{\downarrow 2}}{2}\right)$$

and substituting for ϵ_F^σ, we get

$$E_{\text{kin}} = \frac{n^2 + m^2}{4\rho}.$$

Then the total ground state energy of the weakly magnetized ferromagnet can be written as,

$$E_{\text{tot}} = E_0 + \frac{1/\rho - U}{4} m^2.$$

For a general density of states, one expands $\rho(\epsilon)$ near the Fermi level ϵ_F^0 of the paramagnetic metal as,

$$\rho(\epsilon) = \rho(\epsilon_F^0) + (\epsilon - \epsilon_F^0)\rho'(\epsilon_F^0) + \frac{1}{2}(\epsilon - \epsilon_F^0)^2 \rho''(\epsilon_F^0) + \cdots.$$

This is going to introduce higher order terms in the expansion of the total ground state energy as a function of the magnetization, i.e.,

$$E_{\text{tot}} = E_0 + \frac{1/\rho - U}{4} m^2 + b_0 m^4 + O(m^6), \tag{2.14}$$

where the coefficient b_0 depends on the first and second derivatives of the density of states at the Fermi level. Here, we assume it is a positive constant. Notice that only terms even in the magnetization m can appear in this expansion due to the rotational invariance of the Hamiltonian.

The ground state magnetization is that which minimizes the energy. Differentiating equation (2.14) with respect to m and making this equal to zero, we obtain two solutions,

$$m = \begin{cases} 0, & \text{for} \quad U < U_c \\ \sqrt{\dfrac{U - U_c}{8b_0}}, & \text{for} \quad U > U_c \end{cases} .$$

where $U_c = 1/\rho$. Then, in the absence of an external magnetic field, the system requires a critical value of the Coulomb repulsion $U_c = 1/\rho$ for a spontaneous magnetization to appear. For $U < U_c$ the metal is in a paramagnetic state even at $T = 0$. At zero temperature there is a paramagnetic-to-ferromagnetic metal phase transition driven by the strength of the Coulomb repulsion U. The magnetization vanishes near U_c as $m \propto |U - U_c|^{\beta}$ with the critical exponent $\beta = 1/2$.

If we add to the system a magnetic field h in the z-direction, there is an extra Zeemann term in the total energy, $E_{\text{Zeemann}} = -hm$. In this case its minimization yields,

$$m = \frac{h}{2(U - U_c) + 4b_0 m^2}. \tag{2.15}$$

The magnetic susceptibility $\chi = \partial m/\partial h$ is given by,

$$\chi = \frac{1}{2(U - U_c) + 12b_0 m^2}.$$

The susceptibility diverges on both sides of the ferromagnetic transition as $\chi \propto |U - U_c|^{-\gamma}$ with the exponent $\gamma = 1$. Also at the *quantum critical point* $U = U_c$, we have using equation (2.15)

$$m = h^{1/\delta}$$

where the critical exponent $\delta = 3$.

In the Hartree–Fock approximation, the energies of the quasiparticles are $E_k^{\sigma} = \epsilon_k + U\langle n_{-\sigma}\rangle$. Then $E_k^{\downarrow} - E_k^{\uparrow} = Um = \Delta$, where Δ is the *exchange gap* that vanishes at the transition as the magnetization, i.e., $\Delta \propto |U - U_c|^{1/2}$. This is a gap for spin–flip excitations, and is shown schematically in figure 2.2. The metallic ferromagnet, however, supports another type of low energy magnetic excitations, the *spin-waves*, that will be studied below for the case of a ferromagnet with localized magnetic moments.

Up to now our analysis is restricted to zero temperature. However, there are also finite temperature magnetic transitions in the model within the Hartree–Fock, mean-field approximation. This can be seen from the temperature dependence of the occupation numbers shown in equation (2.13). The effect of temperature can be calculated and for small temperatures ($T/\epsilon_F^0 << 1$) and magnetizations it results in a *free energy*,

$$F = F_0(T) + (U - U_c - aT^2)m^2 + b(T)m^4, \tag{2.16}$$

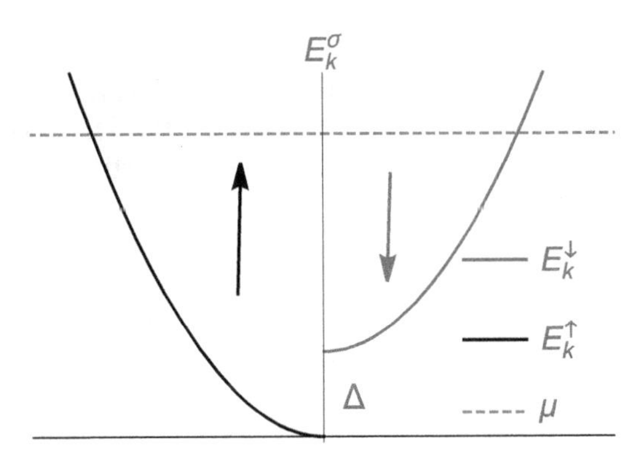

Figure 2.2. Quasi-particle dispersion relations of a ferromagnetic metallic system in the mean-field approximation. Δ is the exchange or Stoner gap. The horizontal axis is the density of states.

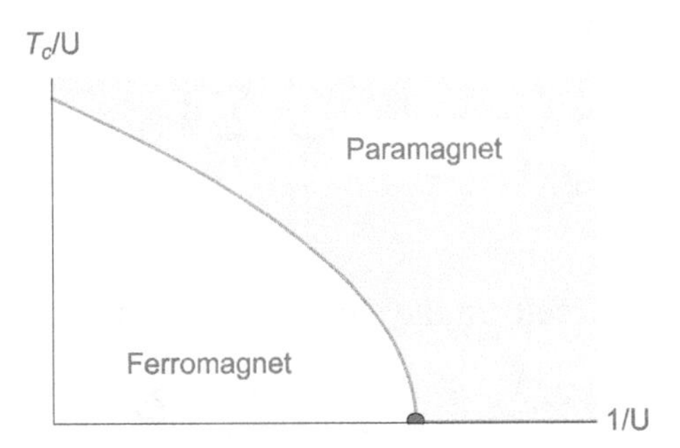

Figure 2.3. Phase diagram of the ferromagnetic Hubbard model in the mean-field approximation. There is a zero temperature transition at U_c and a line of finite temperature transitions, $T_c = \sqrt{(U - U_c)/a}$. In the mean-field approximation there are no fluctuations in the paramagnetic phase. The critical exponents are the same for the quantum and thermal phase transitions.

where F_0 is analytic at the transition, a is a positive constant and $b(T) = b_0 + O((T/\epsilon_F^0)^2)$ where the temperature correction can be safely neglected. Minimizing the free energy with respect to m, we obtain that now the magnetization vanishes at a U dependent critical temperature $T_c(U) = [(U - U_c)/a]^{1/2}$. Near $T_c(U)$ the magnetization vanishes as $m \propto \sqrt{|T - T_c(U)|}$ with the same exponent $\beta = 1/2$ of the zero temperature transition. The phase diagram of the metallic ferromagnet in the mean-field approximation is shown in figure 2.3. From the expression for the free energy, we can obtain the non-analytic part of the specific heat $C_V/T = -\partial^2 F/\partial T^2$ and this presents a jump at the finite temperature transition since it is zero in the paramagnetic phase where $m = 0$. If we describe this behavior using $C_V \propto |T_c(U) - T|^{-\alpha}$, the mean-field jump is associated with a critical exponent $\alpha = 0$. A remarkable feature of the mean-field solution is the absence of fluctuations in the paramagnetic phase of the system (see figure 2.3). Of course for $T = 0$ there is no specific heat and the critical exponent α is defined through the singular part of the ground state energy, i.e., $F_s \propto |U - U_c|^{2-\alpha}$.

Hubbard I approximation
Consider the last Green's function in equation (2.9). Previously we decoupled this propagator as in equation (2.10), which resulted in the Hartree–Fock approximation. In the Hubbard I approximation, we go one step further and write the equation of motion for this propagator,

$$(\omega - U)\left\langle\left\langle n_{i-\sigma}c_{i\sigma}; c_{j\sigma}^{+}\right\rangle\right\rangle_{\omega} = \langle n_{i-\sigma}\rangle\delta_{ij} + \sum_{l} t_{il}\left\langle\left\langle n_{i-\sigma}c_{l\sigma}; c_{j\sigma}^{+}\right\rangle\right\rangle_{\omega} + \sum_{l} t_{il}\left\langle\left\langle (c_{i-\sigma}^{+}c_{l-\sigma} - c_{l-\sigma}^{+}c_{i-\sigma})c_{l\sigma}; c_{j\sigma}^{+}\right\rangle\right\rangle_{\omega}. \tag{2.17}$$

The Green's function

$$\left\langle\left\langle n_{i-\sigma}c_{l\sigma}; c_{j\sigma}^{+}\right\rangle\right\rangle_{\omega} \approx \langle n_{i-\sigma}\rangle\left\langle\left\langle c_{l\sigma}; c_{j\sigma}^{+}\right\rangle\right\rangle_{\omega} \approx \langle n_{-\sigma}\rangle\left\langle\left\langle c_{l\sigma}; c_{j\sigma}^{+}\right\rangle\right\rangle_{\omega} \tag{2.18}$$

is decoupled as before, and

$$\sum_{l} t_{il}\left\langle\left\langle (c_{i-\sigma}^{+}c_{l-\sigma} - c_{l-\sigma}^{+}c_{i-\sigma})c_{l\sigma}; c_{j\sigma}^{+}\right\rangle\right\rangle_{\omega} \approx \sum_{l} t_{il}\langle c_{i-\sigma}^{+}c_{l-\sigma} - c_{l-\sigma}^{+}c_{i-\sigma}\rangle \times \left\langle\left\langle c_{l\sigma}; c_{j\sigma}^{+}\right\rangle\right\rangle_{\omega} = 0. \tag{2.19}$$

Substituting the last two results in the equation of motion, equation (2.17), we get,

$$(\omega - U)\left\langle\left\langle n_{i-\sigma}c_{i\sigma}; c_{j\sigma}^{+}\right\rangle\right\rangle_{\omega} = \langle n_{-\sigma}\rangle\delta_{ij} + \langle n_{-\sigma}\rangle\sum_{l} t_{il}\left\langle\left\langle c_{l\sigma}; c_{j\sigma}^{+}\right\rangle\right\rangle_{\omega}. \tag{2.20}$$

Substituting that in equation (2.9) and Fourier transforming in lattice space yields

$$\left(\omega - \epsilon_k\left(1 + \frac{U}{\omega - U}\langle n_{-\sigma}\rangle\right)\right)\langle\langle c_{k\sigma}; c_{k'\sigma}^{+}\rangle\rangle_{\omega} = \delta_{kk'}\left(1 + \frac{U}{\omega - U}\langle n_{-\sigma}\rangle\right). \tag{2.21}$$

In the limit $U \to \infty$, of very large Coulomb repulsion, we obtain

$$\langle\langle c_{k\sigma}; c_{k'\sigma}^{+}\rangle\rangle_{\omega} = \delta_{kk'}\frac{(1 - \langle n_{-\sigma}\rangle)}{\omega - \epsilon_k(1 - \langle n_{-\sigma}\rangle)}, \tag{2.22}$$

so that the band of spin σ is *narrowed* by the occupation of the band of opposite spin.

The Green's function in equation (2.21) has two poles as a function of frequency, which give rise to two *Hubbard bands.*

Although the original purpose of the Hubbard model was to describe magnetic solutions, it also describes a metal-insulator transition. This can be seen from the two solutions of equation (2.21). For a band of total width W, we can easily check the existence of a paramagnetic solution for a total occupation of one electron per site, i.e., $\langle n_{\uparrow}\rangle = \langle n_{\downarrow}\rangle = 1/2$. The difference between the bottom of the upper Hubbard E_b^u band (UHB) and the top of the lower Hubbard band (LHB) E_t^l is given by,

$$E_b^u - E_t^l = \sqrt{U^2 + (W/2)^2} - W/2,$$

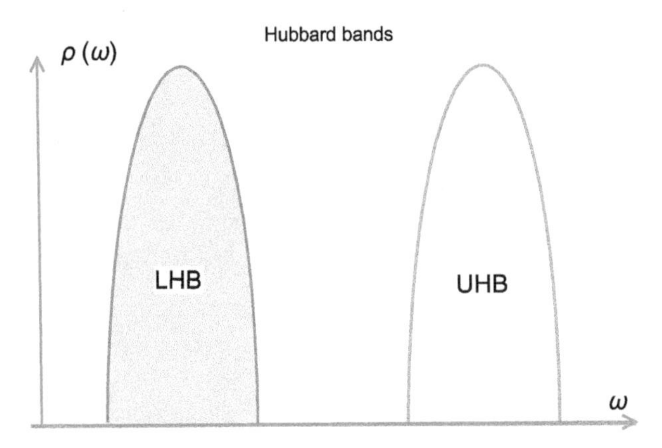

Figure 2.4. Lower (LHB) and upper (UHB) Hubbard bands. The LHB of singly occupied states can accommodate one electron per site. Then for $\langle n \rangle = 1$ and $U > 0$ there is a gap between the bands and the system is a Mott insulator, since the Fermi level is in the gap (schematic).

which is always positive for $U > 0$ (see figure 2.4). Consequently, the two Hubbard bands split for $U > U_c = 0$ and since the LHB can accommodate one electron per site, the ground state for $\langle n_\uparrow \rangle + \langle n_\downarrow \rangle = 1$ is a *Mott insulator*. In the case of the Hubbard I approximation the Mott transition occurs for $U = 0$. This is not necessary and other methods or approximations yield a more realistic results that require a finite value of U_c to give rise to a Mott transition.

2.1.4 BCS superconductivity

The Bardeen, Cooper, Schrieffer (BCS) theory of superconductivity can be obtained from an attractive Hubbard model using the Green's function approach. The Hamiltonian is given by,

$$H = \sum_{ij\sigma} t_{ij} c_{i\sigma}^{+} c_{j\sigma} - \frac{U}{2} \sum_{i\sigma} n_{i\sigma} n_{i-\sigma}, \tag{2.23}$$

where $U > 0$ is an attractive interaction that in the BCS theory is due to the electron–phonon coupling. The superconductivity order parameter is given by

$$\Delta = \frac{1}{N} \sum_{k\sigma} \langle c_{k\sigma}^{+} c_{-k-\sigma}^{+} \rangle = \frac{1}{N} \sum_{i\sigma} \langle c_{i\sigma}^{+} c_{i-\sigma}^{+} \rangle, \tag{2.24}$$

and $\Delta^* = (1/N)\sum_{i\sigma} \langle c_{i-\sigma} c_{i\sigma} \rangle$, where N is the number of lattice sites. At $T = 0$, the expectation value is taken in the BCS ground state that does not conserve the total number of particles. The order parameter can be obtained from the appropriate Green's function, $\langle\langle c_{i\sigma}^{+}; c_{j\sigma'}^{+} \rangle\rangle$ using the leap theorem. This is an anomalous Green's function first introduced by Gorkov. It obeys the following equation of motion

$$\omega \left\langle \left\langle c_{i\sigma}^{+}; c_{j\sigma'}^{+} \right\rangle \right\rangle = \frac{1}{2\pi} \left\langle \left\{ c_{i\sigma}^{+}, c_{j\sigma'}^{+} \right\} \right\rangle + \left\langle \left\langle \left[c_{i\sigma}^{+}, H \right]; c_{j\sigma'} \right\rangle \right\rangle. \tag{2.25}$$

In this case the anticommutator vanishes and calculating the commutator of $c_{i\sigma}^{+}$ with the Hamiltonian equation (2.23), we obtain

$$\omega\Big\langle\Big\langle c_{i\sigma}^{+};\, c_{j\sigma'}^{+}\Big\rangle\Big\rangle = -\sum_{l\sigma} t_{il}\Big\langle\Big\langle c_{l\sigma}^{+};\, c_{j\sigma'}^{+}\Big\rangle\Big\rangle + U\Big\langle\Big\langle c_{i\sigma}^{+}c_{i-\sigma}^{+}c_{i-\sigma};\, c_{j\sigma'}^{+}\Big\rangle\Big\rangle, \tag{2.26}$$

which generates a higher order Green's function due to the many-body interaction U. The BCS approximation consists of the following decoupling of this propagator,

$$\Big\langle\Big\langle c_{i\sigma}^{+}c_{i-\sigma}^{+}c_{i-\sigma};\, c_{j-\sigma}^{+}\Big\rangle\Big\rangle \approx \langle c_{i\sigma}^{+}c_{i-\sigma}^{+}\rangle\Big\langle\Big\langle c_{i-\sigma};\, c_{j\sigma'}^{+}\Big\rangle\Big\rangle + \langle c_{i-\sigma}^{+}c_{i-\sigma}\rangle\Big\langle\Big\langle c_{i\sigma}^{+};\, c_{j\sigma'}^{+}\Big\rangle\Big\rangle,$$

such that the equation of motion becomes

$$\begin{aligned}\omega\Big\langle\Big\langle c_{i\sigma}^{+};\, c_{j\sigma'}^{+}\Big\rangle\Big\rangle = \sum_{l\sigma} t_{il}\Big\langle\Big\langle c_{l\sigma}^{+};\, c_{j\sigma'}^{+}\Big\rangle\Big\rangle + U\langle c_{i\sigma}^{+}c_{i-\sigma}^{+}\rangle\Big\langle\Big\langle c_{i-\sigma};\, c_{j\sigma'}^{+}\Big\rangle\Big\rangle \\ + U\langle n_{i-\sigma}\rangle\Big\langle\Big\langle c_{i\sigma}^{+};\, c_{j\sigma'}^{+}\Big\rangle\Big\rangle\end{aligned} \tag{2.27}$$

with $\langle c_{i-\sigma}^{+}c_{i-\sigma}\rangle = \langle n_{i-\sigma}\rangle = \langle n_{-\sigma}\rangle$, where we used translation invariance.

A new Green's function $\langle\langle c_{i-\sigma};\, c_{j\sigma'}^{+}\rangle\rangle$ has been generated in spite of the decoupling. This obeys the following equation of motion

$$\omega\Big\langle\Big\langle c_{i-\sigma};\, c_{j\sigma'}^{+}\Big\rangle\Big\rangle = \delta_{ij}\delta_{\sigma'-\sigma} + \sum_{il} t_{il}\Big\langle\Big\langle c_{l-\sigma};\, c_{j\sigma'}^{+}\Big\rangle\Big\rangle - U\Big\langle\Big\langle c_{i\sigma}^{+}c_{i\sigma}c_{i-\sigma};\, c_{j\sigma'}^{+}\Big\rangle\Big\rangle \tag{2.28}$$

The last higher order Green's function is decoupled in the same spirit as before,

$$\Big\langle\Big\langle c_{i\sigma}^{+}c_{i\sigma}c_{i-\sigma};\, c_{j\sigma'}^{+}\Big\rangle\Big\rangle \approx -\langle c_{i-\sigma}c_{i\sigma}\rangle\Big\langle\Big\langle c_{i\sigma}^{+};\, c_{j\sigma'}^{+}\Big\rangle\Big\rangle + \langle n_{\sigma}\rangle\Big\langle\Big\langle c_{i-\sigma};\, c_{j\sigma'}^{+}\Big\rangle\Big\rangle, \tag{2.29}$$

such that equation (2.28) can be written as

$$\begin{aligned}\omega\Big\langle\Big\langle c_{i-\sigma};\, c_{j\sigma'}^{+}\Big\rangle\Big\rangle = \delta_{ij}\delta_{\sigma'-\sigma} + \sum_{il} t_{il}\Big\langle\Big\langle c_{l-\sigma};\, c_{j\sigma'}^{+}\Big\rangle\Big\rangle + U\langle c_{i-\sigma}c_{i\sigma}\rangle\Big\langle\Big\langle c_{i\sigma}^{+};\, c_{j\sigma'}^{+}\Big\rangle\Big\rangle \\ - U\langle n_{\sigma}\rangle\Big\langle\Big\langle c_{i-\sigma};\, c_{j\sigma'}^{+}\Big\rangle\Big\rangle\end{aligned} \tag{2.30}$$

Finally Fourier transforming in lattice space ($\times\sum_{ij} e^{ik\cdot r_i}e^{-ik'\cdot r_j}$), we obtain for equations (2.27) and (2.30), respectively,

$$\begin{aligned}\omega\langle\langle c_{-k\sigma}^{+};\, c_{k'\sigma'}^{+}\rangle\rangle = -\,\epsilon_k\langle\langle c_{-k\sigma}^{+};\, c_{k'\sigma'}^{+}\rangle\rangle + U\Delta\langle\langle c_{k-\sigma};\, c_{k'\sigma'}^{+}\rangle\rangle \\ + U\langle n_{i-\sigma}\rangle\langle\langle c_{k\sigma}^{+};\, c_{k'\sigma'}^{+}\rangle\rangle,\end{aligned}$$

and

$$\begin{aligned}\omega\langle\langle c_{k-\sigma};\, c_{k'\sigma'}^{+}\rangle\rangle = \delta_{kk'}\delta_{-\sigma\sigma'} + \epsilon_k\langle\langle c_{k-\sigma};\, c_{k'\sigma'}^{+}\rangle\rangle + U\Delta'\langle\langle c_{-k\sigma}^{+};\, c_{k'\sigma'}^{+}\rangle\rangle \\ -\, U\langle n_{\sigma}\rangle\langle\langle c_{k-\sigma};\, c_{k'\sigma'}^{+}\rangle\rangle,\end{aligned}$$

where

$$\Delta = \langle c_{i-\sigma}c_{i\sigma}\rangle$$

and

$$\Delta' = \langle c_{i\sigma}^{+} c_{i-\sigma}^{+} \rangle$$

are both site independent due to translation invariance. These equations yield,

$$(\omega + \epsilon_k - U\langle n_{-\sigma}\rangle)\langle\langle c_{-k\sigma}^{+}; c_{k'\sigma'}^{+}\rangle\rangle = U\Delta\langle\langle c_{k-\sigma}; c_{k'\sigma'}^{+}\rangle\rangle, \tag{2.31}$$

and

$$(\omega - \epsilon_k + U\langle n_\sigma\rangle)\langle\langle c_{k-\sigma}; c_{k'\sigma'}^{+}\rangle\rangle = \delta_{kk'}\delta_{-\sigma\sigma'} + U\Delta'\langle\langle c_{-k\sigma}^{+}; c_{k'\sigma'}^{+}\rangle\rangle. \tag{2.32}$$

Solving equation (2.32) for $\langle\langle c_{k-\sigma}; c_{k'\sigma'}^{+}\rangle\rangle$, substituting in equation (2.31) and using that $\Delta' = \Delta^*$ and $\langle n_\sigma\rangle = \langle n_{-\sigma}\rangle$ since the system is paramagnetic, we get

$$(\omega + E_k)\langle\langle c_{-k\sigma}^{+}; c_{k'\sigma'}^{+}\rangle\rangle = U\Delta\left(\frac{\delta_{kk'}\delta_{-\sigma\sigma'} + U\Delta^*\langle\langle c_{-k\sigma}^{+}; c_{k'\sigma'}^{+}\rangle\rangle}{\omega - E_k}\right), \tag{2.33}$$

where $E_K = \epsilon_k - (U/2)\langle n\rangle$ and $n = \langle n_\sigma\rangle + \langle n_{-\sigma}\rangle$. Equation (2.33) can be rewritten as,

$$\left(\omega + E_k - \frac{U|\Delta|^2}{\omega - E_k}\right)\langle\langle c_{-k\sigma}^{+}; c_{k'\sigma'}^{+}\rangle\rangle = U\Delta\left(\frac{\delta_{kk'}\delta_{-\sigma\sigma'}}{\omega - E_k}\right). \tag{2.34}$$

Finally, we get for the anomalous Green's function

$$\langle\langle c_{-k\sigma}^{+}; c_{k'\sigma'}^{+}\rangle\rangle = \frac{U\Delta}{(\omega + E_k)(\omega - E_k) - U^2|\Delta|^2}\delta_{kk'}\delta_{-\sigma\sigma'}, \tag{2.35}$$

or

$$\langle\langle c_{k\sigma}^{+}; c_{-k-\sigma}^{+}\rangle\rangle = \frac{U\Delta}{(\omega + E_k)(\omega - E_k) - U^2|\Delta|^2}, \tag{2.36}$$

where we exchanged $k \to -k$. For the normal Green's function we obtain

$$\langle\langle c_{k-\sigma}; c_{k'\sigma'}^{+}\rangle\rangle = \frac{\delta_{kk'}\delta_{-\sigma\sigma'}}{(\omega - E_k)}\left(1 + \frac{U^2|\Delta|^2}{(\omega + E_k)(\omega - E_k) - U^2|\Delta|^2}\right) \tag{2.37}$$

or

$$\langle\langle c_{k\sigma'}; c_{k\sigma'}^{+}\rangle\rangle = \frac{\omega + E_k}{(\omega + E_k)(\omega - E_k) - U^2|\Delta|^2}. \tag{2.38}$$

Notice that both Green's functions have the same poles,

$$\omega_{1,2}(k) = \pm\sqrt{\epsilon_k^2 + U^2|\Delta|^2}, \tag{2.39}$$

where we replaced $E_k \to \epsilon_k$, since the term $-(U/2)\langle n\rangle$ is a constant shift of the band and can be neglected. These energies correspond to the excitations of the superconductor that become gapped due to the pairing of electrons in this new state. In order to obtain the order parameter Δ, we write,

$$\langle\langle c_{k\sigma}^{+}; c_{-k-\sigma}^{+}\rangle\rangle = \frac{U\Delta}{(\omega - \omega_1)(\omega - \omega_2)} = \frac{U\Delta}{2\omega_1}\left(\frac{1}{\omega - \omega_1} - \frac{1}{\omega + \omega_1}\right) \tag{2.40}$$

since $\omega_2(k) = -\omega_1(k)$. Using the leap theorem, we obtain,

$$\sum_k \langle c_{-k-\sigma}^{+} c_{k\sigma}^{+}\rangle = U\Delta \sum_k \frac{\tanh(\beta\omega(k)/2)}{2\omega(k)}, \tag{2.41}$$

where $\omega(k) = \omega_1(k) = \sqrt{\epsilon_k^2 + |\tilde{\Delta}|^2}$, with $\tilde{\Delta} = U\Delta$. The *gap equation* can finally be written as,

$$\frac{1}{U} = \sum_k \frac{\tanh(\beta\omega(k)/2)}{2\omega(k)}. \tag{2.42}$$

For convenience, we replace the sum in k by an energy integration. For this purpose we use that $\sum_k f(\epsilon_k) = \int d\epsilon \sum_k \delta(\epsilon - \epsilon_k) f(\epsilon)$ and introduce the density of electronic states $\rho(\epsilon) = \sum_k \delta(\epsilon - \epsilon_k)$. In the BCS approximation, we integrate in a small energy sphere of width $2\hbar\omega_D$ around the Fermi energy, ϵ_F. In this small energy interval, which is determined by the Debye temperature T_D ($k_B T_D = \hbar\omega_D$, a small energy scale compared to the bandwidth) the density of states can be taken as constant and given by $\rho(\epsilon_F)$. Then we obtain the usual form for the BCS gap equation,

$$\frac{1}{U\rho(\epsilon_F)} = \int_{-\hbar\omega_D}^{\hbar\omega_D} d\epsilon \frac{\tanh(\beta\omega(\epsilon, \tilde{\Delta})/2)}{2\omega(\epsilon, \tilde{\Delta})}. \tag{2.43}$$

The normalized solution of this equation is shown in figure 2.5. The zero temperature order parameter can be easily obtained from equation (2.5). Since $\tanh(x \to \infty) \approx 1$, we get,

$$\frac{1}{U\rho(\epsilon_F)} = \int_0^{\hbar\omega_D} d\epsilon \frac{1}{\sqrt{\epsilon^2 + \tilde{\Delta}_0^2}} \tag{2.44}$$

This can be easily integrated to yield,

$$\frac{1}{U\rho(\epsilon_F)} = \text{arcsinh}\left(\frac{\hbar\omega_D}{\Delta_0}\right)$$

In real situations $\Delta_0 \ll \hbar\omega_D$ and expanding the arcsinh(x) for small x, we get,

$$\boxed{\Delta_0 = 2\hbar\omega_D e^{-\frac{1}{U\rho(\epsilon_F)}}.} \tag{2.45}$$

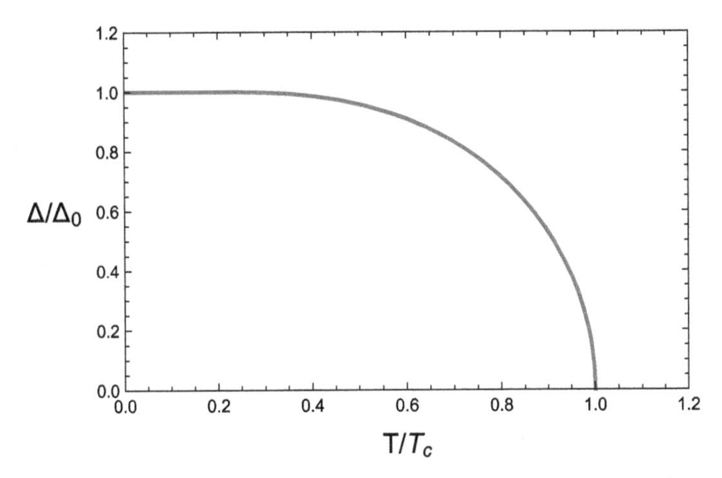

Figure 2.5. Solution of the BCS gap equation, equation (2.43), showing the superconducting order parameter normalized by its value at zero temperature versus temperature normalized by the critical temperature T_c.

The critical temperature T_c can be obtained from equation (2.43) noting that $\tilde{\Delta}(T = T_c) = 0$. This gives the equation,

$$\frac{1}{U\rho(\epsilon_F)} = \int_0^{\hbar\omega_D} d\epsilon \frac{\tanh(\epsilon/2k_B T_c)}{\epsilon}.$$

Now the integration has to be performed by parts and using that $\hbar\omega_D \gg k_B T_c$, we obtain the well-known result,

$$k_B T_c = \frac{2e^{\gamma_E}\hbar\omega_D}{\pi} e^{-\frac{1}{U\rho(\epsilon_F)}}, \tag{2.46}$$

where $\gamma_E \approx 0.577$ is the Euler constant. The ratio,

$$\frac{\Delta_0}{k_B T_c} \approx 1.76.$$

We remark that near the critical temperature the BCS order parameter vanishes as $\tilde{\Delta} \propto (T - T_c)^{1/2}$ as expected due to the mean-field character of this theory.

The appearance of a superconducting state is associated with the breakdown of gauge invariance. This can be seen by multiplying the second quantization annihilation operator by a global phase $e^{i\phi}$ and the creation operator by its conjugate $e^{-i\phi}$. Normal metals or magnetic systems characterized by occupations numbers $\langle c_{i\sigma}^{+} c_{j\sigma} \rangle$ remain invariant under this operation. However, the superconducting order parameter given by the anomalous correlation function $\langle c_{i\sigma}^{+} c_{j-\sigma}^{+} \rangle$ does not, signaling the breakdown of gauge invariance of the original Hamiltonian by the superfluid state.

2.2 Spin-waves in ferromagnets

In this section we use the Green's function method to obtain the elementary excitations and the magnetization curve of a local moment ferromagnet. In this case, the moments responsible for the magnetism are localized on the lattice sites, differently from those of the transition metals that are itinerant. This localized approach describes the magnetism of insulating materials and is useful also for the case of metallic rare-earths. The most common description of localized magnetism is given by the Heisenberg Hamiltonian,

$$H = -\sum_{ij} J_{ij}\mathbf{S}_i \cdot \mathbf{S}_j,$$

where J_{ij} are taken as nearest neighbor interactions. In the ferromagnetic case $J_{ij} > 0$ and the ground state of the system for $d > 2$ is a ferromagnet with all *quantum* spins $\vec{S}_i$ aligned. The elementary excitations in this state are propagating spin-waves, transverse to the spontaneous magnetization that we assume is in the z-direction.

For spin 1/2 we have the relation,

$$\langle S^z \rangle = S - \langle S_i^- S_i^+ \rangle, \tag{2.47}$$

where $S^{\pm} = S^x \pm iS^y$, such that $\langle S_i^- S_j^+ \rangle$ is the correlation function of the transverse excitations. In order to obtain this correlation function and calculate the magnetization $\langle S_z \rangle$ using equation (2.47) we consider the Green's function $\langle\langle S_j^+; S_i^- \rangle\rangle$ that obeys the following equation of motion,

$$\omega\left\langle\left\langle S_i^+; S_j^- \right\rangle\right\rangle = \left\langle\left[S_i^+, S_j^-\right]\right\rangle + \left\langle\left\langle [S_i^+, H]; S_j^- \right\rangle\right\rangle. \tag{2.48}$$

Since $\mathbf{S}_i \cdot \mathbf{S}_j = S_i^z S_j^z + (1/2)(S_i^+ S_j^- + S_j^- S_i^+)$ the commutator of S_j^+ with H can be easily calculated. Using the commutation relations for the spin operators, $[S^+, S^-] = 2S^z$; $[S^z, S^+] = S^+$ and $[S^z, S^-] = -S^-$, we get,

$$[S_i^{\pm}, H] = \pm\sum_l J_{il}(S_l^z S_i^{\pm} - S_i^z S_l^{\pm})$$

and the equation of motion for the transverse Green's function becomes,

$$\begin{aligned}\omega\left\langle\left\langle S_i^+; S_j^- \right\rangle\right\rangle = 2\langle S_i^z \rangle\delta_{ij} &+ \sum_l J_{il}\left\langle\left\langle S_l^z S_i^+; S_j^- \right\rangle\right\rangle \\ &- \sum_l J_{il}\left\langle\left\langle S_i^z S_l^+; S_j^- \right\rangle\right\rangle.\end{aligned} \tag{2.49}$$

As expected, a new higher order Green's function has been generated. Instead of obtaining its equation of motion we use a decoupling, known as the *random phase approximation* (RPA) that consists in the following,

$$\left\langle\left\langle S_i^z S_l^+; S_j^- \right\rangle\right\rangle \approx \langle S_i^z \rangle\left\langle\left\langle S_l^+; S_j^- \right\rangle\right\rangle = \langle S^z \rangle\left\langle\left\langle S_l^+; S_j^- \right\rangle\right\rangle,$$

where in the last step we used translation invariance. Now the equation of motion can be solved making a Fourier transformation for momentum space. We obtain

$$\omega\langle\langle S_k^+; S_{k'}^-\rangle\rangle = 2\langle S^z\rangle\delta_{kk'} + J(0)\langle S^z\rangle\langle\langle S_k^+; S_{k'}^-\rangle\rangle \\ - J(k)\langle S^z\rangle\langle\langle S_k^+; S_{k'}^-\rangle\rangle, \tag{2.50}$$

where $J(k) = \sum_l J_{il} e^{ik\cdot(r_l - r_i)}$. Finally,

$$\langle\langle S_k^+; S_{k'}^-\rangle\rangle = \frac{2\langle S^z\rangle}{\omega - \langle S^z\rangle(J(0) - J(k))}\delta_{kk'}. \tag{2.51}$$

The poles of this Green's function at $\omega(k) = \langle S^z\rangle(J(0) - J(k))$ give the energies of the spin-wave excitations in the ferromagnet. For a hypercubic lattice, in the *hydrodynamic limit*, i.e., in the limit the $k \to 0$, we get $\omega(k) = D\langle S^z\rangle k^2$. The spin-waves or magnons have a quadratic dispersion and their *stiffness* $D\langle S^z\rangle$ vanishes at the critical temperature T_c, as the magnetization of the system in this RPA approximation. For nearest neighbor exchange interactions of intensity J, the stiffness $D \propto J$.

Notice that the energy of the magnon with $k = 0$ vanishes ($\omega(k = 0) = 0$). The magnon with zero energy is a *Goldstone mode* and corresponds to a uniform rotation of the whole system. Since it costs no energy, it restores the spherical symmetry of the Heisenberg Hamiltonian broken in the ferromagnetic state by the appearance of a spontaneous magnetization.

The magnetization of the system with spins 1/2 can be derived using equation (2.47), and the correlation function $\langle S^- S^+\rangle$ obtained using the Green's function $\langle\langle S_k^+; S_{k'}^-\rangle\rangle$ and the leap theorem. Since spin-waves obey Bose–Einstein statistics, we get for the average magnetization per spin,

$$\langle S^z\rangle = \frac{1}{2} - \frac{1}{(2\pi)^3}\int d^3k \frac{2\langle S^z\rangle}{e^{\beta D\langle S^z\rangle k^2} - 1}. \tag{2.52}$$

At very low temperatures, we can assume $\langle S^z\rangle \sim S$ on the right hand side of equation (2.52) and a simple change of variables shows that the decrease of the magnetization at these temperatures is given by,

$$\langle S^z\rangle \approx \frac{1}{2} - BT^{3/2}.$$

where B is a constant. This is the celebrated Bloch's $T^{3/2}$ law for the low temperature decrease of the magnetization of isotropic ferromagnets and has been extensively verified experimentally in ferromagnetic materials.

Equation (2.52) is a self-consistent equation for the magnetization and can be rewritten as

$$\langle S^z\rangle \frac{1}{(2\pi)^3}\int d^3k\left(1 + \frac{2}{e^{\beta D\langle S^z\rangle k^2} - 1}\right) = \frac{1}{2}. \tag{2.53}$$

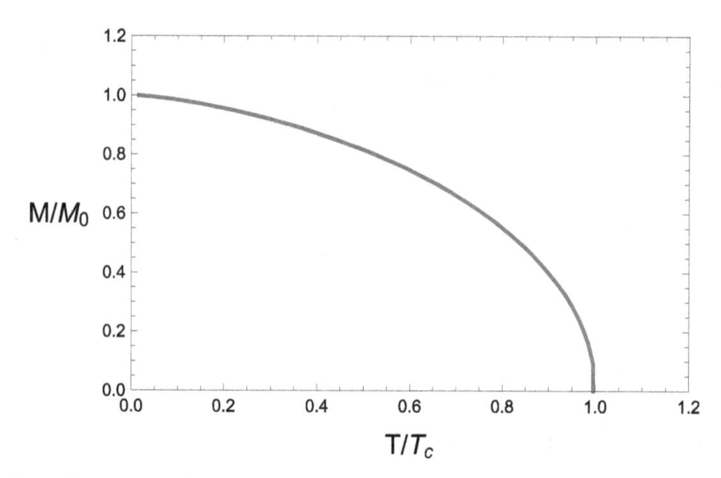

Figure 2.6. Self-consistent, normalized solution of equation (2.54) showing the magnetization versus temperature of a ferromagnet with local moments due to excitations of spin-waves.

For an isotropic three-dimensional (3d) ferromagnet this becomes,

$$\frac{1}{(2\pi)^3}\int 4\pi k^2 dk \coth\frac{\beta D\langle S^z\rangle k^2}{2} = \frac{1}{2\langle S^z\rangle}. \tag{2.54}$$

Close to the critical temperature T_c of the ferromagnet, the magnetization becomes very small. Since $\coth x \approx 1/x$ when $x \to 0$, we obtain $1/\beta_c = k_B T_c = \pi D/4$, where as pointed out before, the constant D is proportional to the nearest neighbor exchange interaction J of the Heisenberg Hamiltonian. The solution of the full self-consistent equation, equation (2.54), is presented in figure 2.6. As in the superconducting case, the magnetization that is the order parameter of the ferromagnetic phase also vanishes with a critical exponent $\beta = 1/2$, i.e., $M = \langle S^z\rangle \propto (T - T_c)^{1/2}$ near T_c. The reason is that the RPA is also a mean-field approximation.

It is interesting to notice that in two dimensions (2d) the integral that gives the number of excited magnons for a non-zero temperature becomes of the form $\int (k/k^2)dk = \int (dk/k)$ that is logarithmic divergent at $k = 0$. Infrared divergencies are very serious in condensed matter, differently from the ultraviolets that we can get rid of considering the discreteness of the lattice that provides a natural cut-off. In 2d, the proliferation of low energy magnons excited for any finite temperature destroys the spontaneous magnetization of the system. This is the core of the Mermin–Wagner theorem that precludes the existence of ferromagnetic long-range order at any finite temperature in isotropic systems in two or fewer dimensions ($d \leqslant 2$) [4].

2.3 Critical exponents

We have studied in this chapter several types of *phase transitions*. At zero temperature, a quantum phase transition from the paramagnetic state to a long-range ordered ferromagnet when the Stoner criterion is satisfied. Also, thermal phase transitions from a superconductor to a normal metal and from a ferromagnet, localized or itinerant to a paramagnetic state, both occurring at some critical temperature T_c. Phase transitions are among the most studied phenomena in condensed matter. They involve the fundamental concept of an order parameter associated with a spontaneously broken symmetry. In the ferromagnetic case, the appearance of a spontaneous magnetization that breaks the rotational symmetry of the Hubbard or of the Heisenberg Hamiltonian. In the superconductor the opening of a gap in the Fermi surface associated with an anomalous correlation function that breaks the gauge invariance of the Hamiltonian. These phase transitions are of *second order*, since the order parameter vanishes continuously, and the non-analytical behavior appears in physical quantities related to second derivatives of the free energy, as the specific heat or susceptibility. The singularities in the physical quantities are characterized by a set of critical exponents, that defines the *universality class* of the transition. In all cases studied in this chapter, we used mean-field approximations and the critical exponents obtained are a direct consequence of this type of approximation.

For completeness we define the critical exponents that are relevant in this book (table 2.1).

The mean-field exponents obtained in this chapter are, $\beta = 1/2$, $\alpha = 0$, $\gamma = 1$, $\delta = 3$. They are the same for the quantum phase transition in the Hubbard model and for the thermal phase transitions in the ferromagnet and superconductor. In this sense they are universal and depend exclusively on the mean-field nature of the approximations we used. In practice, in real systems, the observed critical exponents do not take necessarily these values. Also, in exactly soluble models like the 2d Ising model, the obtained critical exponents are different from the mean-field values. In some cases however, like in the thermal superconducting phase transition they provide a good approximation even very close to the transition.

Table 2.1. The most common critical exponents associated with a second order phase transition. The quantity g is the control parameter, such that the transition occurs at $g = 0$. For thermal phase transitions $g \propto (T - T_c)$. For zero temperature quantum phase transitions g is a function of competing parameters in the Hamiltonian. In the ferromagnetic Hubbard model $g = U - U_c$. The critical exponent α for $T = 0$ transitions is defined through the singular part of the free energy. Notice that h is the *conjugate field* of the order parameter.

Critical exponents		
α	$f \propto \lvert g\rvert^{2-\alpha}$	free energy
α	$C_V \propto \lvert g\rvert^{-\alpha}$	specific heat
β	$m \propto \lvert g\rvert^{\beta}$	order parameter
γ	$\chi \propto \partial^2 f/\partial h^2 \propto \lvert g\rvert^{-\gamma}$	susceptibility
δ	$m \propto h^{1/\delta}$	at $g = 0$
ν	$\xi \propto \lvert g\rvert^{-\nu}$	correlation length
z	$\tau \propto \xi^{z}$	correlation time

A great challenge in condensed matter and statistical mechanics is the calculation of critical exponents. This is accomplished by the renormalization group. In the second part of this book we will present the real space version of this approach so that it can be fully appreciated.

References

[1] Tyablikov S V 1967 *Methods in the Quantum Theory of Magnetism* (New York: Plenum Press)
[2] Hubbard J 1963 *Proc. R. Soc. Lond.* 276 238
[3] Nozieres P 1986 *Magnétisme et Localisation dans les Liquides de Fermi* (Cours Collége de France) unpublished
[4] Anderson P W 1997 *Concepts in Solids, Lectures on the Theory of Solids* (Lecture Notes in Physics vol 58) (Singapore: World Scientific)

IOP Publishing

Key Methods and Concepts in Condensed Matter Physics

Green's functions and real space renormalization group

Mucio A Continentino

Chapter 3

Perturbative method for Green's functions

3.1 Time independent perturbations

In this chapter we present a perturbative approach for the Green's function method [1, 2] starting with time independent perturbations. For the purpose of presenting the method we consider the general problem of an instability to a magnetic state characterized by an arbitrary wavevector **q** in a correlated metal. We obtain a generalized Stoner criterion and the wavevector **q** for which this is first satisfied determines the magnetic order in the metal, ferromagnetic, antiferromagnetic or helicoidal.

3.1.1 Generalized Stoner criterion

The metallic system is described by the single band Hubbard Hamiltonian,

$$H = \sum_{i,j,\sigma} t_{ij} c_{i\sigma}^{+} c_{j\sigma} + U \sum_{i} n_{i\uparrow} n_{i\downarrow}, \tag{3.1}$$

where $n_{i\sigma} = c_{i\sigma}^{+} c_{i\sigma}$. The perturbation is a staggered field that couples to the magnetic degrees of freedom of the system.

$$H_{\text{ext}} = h \sum_{i,\sigma} \sigma n_{i\sigma} e^{-i\mathbf{q}\cdot\mathbf{r}_i}, \tag{3.2}$$

where **q** is the wavevector of the external field and $\sigma = \pm(1/2)$. The relevant Green's function to obtain the polarization of the metal is $g_{ij}(\omega) = \left\langle\left\langle c_{i\sigma}; c_{j\sigma}^{+}\right\rangle\right\rangle_{\omega}$. This is decomposed in two parts,

$$g_{ij}(\omega) = g_{ij}^{0}(\omega) + g_{ij}^{1}(\omega),$$

where the first is the zero order Green's function in the absence of the perturbation and the second, the first order modified propagator due to the external staggered field.

doi:10.1088/978-0-7503-3395-5ch3

The zero order propagators obey the equation of motion

$$\omega g_{ij}^{0}(\omega) = \delta_{ij} + \sum_{l} t_{il} g_{lj}^{0} + U\langle n_{-\sigma}\rangle g_{ij}^{0}(\omega),$$

where we used the Hartree–Fock decoupling scheme. We consider the system is in the paramagnetic phase, either because the temperature is too high, i.e., above the Curie temperature or, at zero temperature, because the interaction is too weak and the *Stoner criterion* is not satisfied. In this case, in the absence of the external perturbation h, we have $\langle n_{\uparrow}\rangle = \langle n_{\downarrow}\rangle = n/2$.

The first order Green's function obeys the following equation of motion

$$\omega g_{ij}^{1}(\omega) = \sum_{l} t_{il} g_{lj}^{1}(\omega) + U\langle n_{-\sigma}\rangle g_{ij}^{1}(\omega) + h\sigma e^{-i\mathbf{q}\cdot\mathbf{r}_i} g_{ij}^{0}(\omega)$$
$$+ U\Delta n_{i-\sigma} g_{ij}^{0}(\omega),$$

where $n_{-\sigma}$ and $\Delta n_{i-\sigma}$ are the number of electrons of spin σ in the absence and in the presence of the perturbation, respectively. The latter should be determined by the first order propagators.

Fourier transforming these equations, we have,

$$(\omega - E_k) g_{kk'}^{0}(\omega) = \delta_{k,k'}, \tag{3.3}$$

and

$$(\omega - E_k) g_{kk'}^{1}(\omega) = h\sigma\delta_{k-q,k'} g_{k'}^{0}(\omega) + U\Delta n_{k-k'}^{-\sigma} g_{k'}^{0}(\omega) \tag{3.4}$$

where $E_k = \epsilon_k + Un/2$. Substituting equation (3.3) in equation (3.4), we get

$$(\omega - E_k) g_{kk'}^{1}(\omega) = \frac{1}{\omega - E_{k-q}}(h\sigma + U\Delta n_{k-k'}^{-\sigma}).$$

We now replace $k \to k + q$ and $k' \to k$ to obtain

$$g_{k+q,k}^{1}(\omega) = \frac{1}{(\omega - E_{k+q})}\left(h\sigma + U\Delta n_{q}^{-\sigma}\right)\frac{1}{(\omega - E_k)}.$$

Defining (see equation (2.4))

$$\chi_0(k, q) = \mathcal{F}_{\omega}\left[\frac{1}{(\omega - E_{k+q})}\frac{1}{(\omega - E_k)}\right] \tag{3.5}$$

and using the leap theorem, we can write,

$$\left\langle c_{k\sigma}^{+} c_{k+q\sigma}\right\rangle^{1} = \left[h\sigma + U\Delta n_{q}^{-\sigma}\right]\chi_0(k, q).$$

Since $\Delta n_{q}^{\sigma} = \sum_k \left\langle c_{k\sigma}^{+} c_{k+q\sigma}\right\rangle^{1}$, we get,

$$\Delta n_{q}^{\sigma} = \left[h\sigma + U\Delta n_{q}^{-\sigma}\right]\chi(q)$$

with $\chi(q) = \sum_k \chi_0(k, q)$. Particle number conservation implies $\Delta n_q^{-\sigma} = -\Delta n_q^\sigma$ and defining $\delta m_q = \Delta n_q^\uparrow - \Delta n_q^\downarrow$, where

$$\Delta n_q^\sigma = \frac{\chi(q)}{1 + U\chi(q)} h\sigma,$$

we finally get,

$$\delta m_q = \frac{\chi(q)}{1 + U\chi(q)} h. \tag{3.6}$$

The criterion for a magnetic instability characterized by a wavevector $\mathbf{q}$ is

$$1 + U\chi(q) = 0. \tag{3.7}$$

In this case even in the absence of an external field, the order parameter becomes finite due to the divergence of the susceptibility. The actual wavevector characterizing the type of magnetic order that appears in the system is that for which equation (3.7) is first satisfied. If this occurs for $q = 0$ the system becomes ferromagnetic and for $\mathbf{q} = (\pi, \pi, \pi)$ the instability is to a three-dimensional antiferromagnetic state. For a general q the ground state is a *spin density wave* that may be commensurate or incommensurate with the lattice.

The calculation of the susceptibility $\chi(q)$ is carried out in the following way. First notice that,

$$\frac{1}{(\omega - E_{k+q})} \frac{1}{(\omega - E_k)} = \frac{1}{E_{k+q} - E_k} \left[\frac{1}{\omega - E_{k+q}} - \frac{1}{\omega - E_k} \right].$$

Now we can apply the leap theorem in equation (3.5) to obtain,

$$\chi(q) = \sum_k \chi_0(k, q) = \sum_k \frac{f(E_{k+q}) - f(E_k)}{E_{k+q} - E_k},$$

where $f(E)$ is the Fermi function. In the limit of very low temperatures and for a homogeneous magnetic field $q \to 0$, we obtain $\chi(0) \approx -\rho(\epsilon_F)$, where $\rho(\epsilon_F)$ is the density of states at the Fermi level. In this case the condition for a ferromagnetic instability is given by

$$1 - U\rho(\epsilon_F) = 0. \tag{3.8}$$

This is the Stoner criterion, introduced previously, which is associated with a quantum, zero temperature phase transition from a paramagnetic to a ferromagnetic ground state at the critical value $U_c = 1/\rho(\epsilon_F)$ of the Coulomb repulsion. For finite temperatures the condition,

$$1 - U\chi(q = 0, T_c) = 0$$

defines the critical line $T_c(U)$ of finite temperature ferromagnetic transitions.

3.1.2 Thouless criterion for superconductivity

Let us consider a system described by the attractive Hubbard Hamiltonian, equation (2.23) in the presence of a fictitious field [3] that couples to the superconducting order parameter giving rise to an interaction of the form,

$$H_{\text{int}} = -g \sum_i e^{iq \cdot r_i} \left(c_{i\uparrow}^{+} c_{i\downarrow}^{+} + c_{i\uparrow} c_{i\downarrow} \right). \tag{3.9}$$

Let us assume that temperature is high enough, such that in the absence of the fictitious field the system is a normal metal. We want to calculate the response of the system to the fictitious field in first order perturbation theory. In zero order the relevant Green's functions are

$$\langle\langle c_{k\sigma}; c_{k'\sigma'}^{+} \rangle\rangle_{\omega}^{0} = \frac{\delta_{kk'}\delta_{\sigma\sigma'}}{\omega - \epsilon_k} \tag{3.10}$$

$$\langle\langle c_{k\sigma}^{+}; c_{k'\sigma'}^{+} \rangle\rangle_{\omega}^{0} = 0. \tag{3.11}$$

In first order in g, the equations of motion for the Green's functions are given by,

$$\begin{aligned} \omega\left\langle\left\langle c_{i\sigma}^{+}; c_{j\sigma'}^{+} \right\rangle\right\rangle_{\omega}^{1} = &- \sum_{l\sigma} t_{il} \left\langle\left\langle c_{l\sigma}^{+}; c_{j\sigma'}^{+} \right\rangle\right\rangle_{\omega}^{1} + U\left\langle\left\langle c_{i\sigma}^{+} c_{i-\sigma}^{+} c_{i-\sigma}; c_{j\sigma'}^{+} \right\rangle\right\rangle_{\omega}^{1} \\ &+ g\sigma e^{iq \cdot r_i} \left\langle\left\langle c_{i-\sigma}; c_{j\sigma'}^{+} \right\rangle\right\rangle_{\omega}^{0}. \end{aligned} \tag{3.12}$$

Decoupling the higher order Green's function in the same spirit of the BCS approximation, we have,

$$U\left\langle\left\langle c_{i\sigma}^{+} c_{i-\sigma}^{+} c_{i-\sigma}; c_{j\sigma'}^{+} \right\rangle\right\rangle_{\omega}^{1} \approx U\langle c_{i\sigma}^{+} c_{i-\sigma}^{+} \rangle^{1} \left\langle\left\langle c_{i-\sigma}; c_{j\sigma'}^{+} \right\rangle\right\rangle_{\omega}^{0},$$

where

$$\langle c_{i\sigma}^{+} c_{i-\sigma}^{+} \rangle^{1} = \delta\Delta e^{iq \cdot r_i} \tag{3.13}$$

is of first order in the fictitious field. We can now Fourier transform in space equation (3.12) to obtain

$$\begin{aligned} (\omega + \epsilon_k)\langle\langle c_{k\sigma}^{+}; c_{k'\sigma'}^{+} \rangle\rangle_{\omega}^{1} = &\, U\delta\Delta\left\langle\left\langle c_{k+q,-\sigma}; c_{k'\sigma'}^{+} \right\rangle\right\rangle_{\omega}^{0} \\ &+ g\sigma\left\langle\left\langle c_{k+q,-\sigma}; c_{k'\sigma'}^{+} \right\rangle\right\rangle_{\omega}^{0}, \end{aligned} \tag{3.14}$$

which can be written as

$$\langle\langle c^+_{k\sigma};\, c^+_{k'\sigma'}\rangle\rangle^1_\omega = \frac{U\delta\Delta + g}{(\omega + \epsilon_k)(\omega - \epsilon_{k+q})}. \tag{3.15}$$

Since $\delta\Delta = \sum_k \langle c^+_{k'\sigma'} c^+_{k\sigma}\rangle = \sum_k \mathcal{F}_\omega \langle\langle c^+_{k\sigma};\, c^+_{k'\sigma'}\rangle\rangle^1_\omega$, and defining

$$\chi^0_s(q) = \sum_k \mathcal{F}_\omega \frac{1}{(\omega + \epsilon_k)(\omega - \epsilon_{k+q})}, \tag{3.16}$$

we find

$$\delta\Delta = \frac{\chi^s_0(q)}{1 - U\chi^s_0(q)} g. \tag{3.17}$$

Then the condition for the appearance of superconductivity in the system is given by,

$$1 - U\chi^s_0(q) \tag{3.18}$$

which is known as the *Thouless criterion*. The particle–particle susceptibility can be easily calculated from equation (3.16) and is given by,

$$\chi^s_0 = \sum_k \frac{1 - f(\epsilon_k) - f(\epsilon_{k+q})}{\epsilon_k + \epsilon_{k+q}} \tag{3.19}$$

where we used $f(-\epsilon_k) = 1 - f(\epsilon_k)$. In the limit $q \to 0$, the condition for a homogeneous superconducting ground state is

$$\frac{1}{U} = \sum_k \frac{\tanh(\beta\epsilon_k)/2}{2\epsilon_k}. \tag{3.20}$$

Introducing the density of states $\rho(\omega)$, this can be written as

$$\frac{1}{U} = \frac{1}{2}\int d\omega \rho(\omega) \frac{\tanh \beta\omega/2}{2\omega}.$$

If we now integrate around the Fermi level in an energy interval ω_D that is sufficiently small to take the density of states constant, we get

$$\frac{1}{U} = \rho(\epsilon_F)\int_0^{\omega_D} d\omega \frac{\tanh \beta\omega/2}{\omega}.$$

This finally yields the critical temperature for the superconductor,

$$k_B T_c = 1.14\hbar\omega_D e^{-\frac{1}{U\rho(\epsilon_F)}},$$

which is the BCS critical temperature that we have obtained in chapter 2.

It is also possible for a superconducting instability to occur at a finite wavevector $\mathbf{q}$ giving rise to what is known as a *pair density wave state*. This is the case of a superconductor in a strong external uniform magnetic field where an instability from a normal to a pair density wave ground state is possible when the external magnetic field is sufficiently reduced.

It can be verified from the expression for the particle–particle susceptibility or more directly from T_c, that differently from the ferromagnetic case, in this model the ground state is always superconductor for an arbitrary small value of the attractive interaction (albeit with a small critical temperature T_c), i.e., $U_c = 0$.

3.2 Time dependent perturbations

Let us consider the system described by the Hamiltonian,

$$H = H_0 + V(t)$$

where H_0 is the full Hamiltonian of the system, including the interactions, and $V(t)$ represents a space and time dependent perturbation. This can be, for example, a *field* coupled to the particle density,

$$V(t) = v \sum_j e^{iq \cdot r_j - i\omega t} n_j,$$

where the frequency has an imaginary component, $\omega \to \omega + i\epsilon$, which guarantees that at time $t = -\infty$ the perturbation is turned off. In the previous section the perturbations depended on the position but were time independent. As we have seen in chapter 1, even for time dependent perturbations if the amplitude of the perturbation is sufficiently weak, we can express the response of the system to these perturbations in terms of equilibrium Green's functions. This allows us to extend the results of the previous section for time dependent perturbations [3], as we show below.

3.2.1 The dynamical susceptibility

For simplicity, let us consider as before the case of a paramagnetic metallic system described by the Hubbard Hamiltonian and coupled to a small space and time dependent external magnetic field,

$$H_{\text{ext}} = h \sum_{i,\sigma} \sigma n_{i\sigma} e^{-i\mathbf{q} \cdot \mathbf{r}_i - i\omega t}, \tag{3.21}$$

where the frequency ω has a positive imaginary part. The Green's functions are again separated in zero order and first order contributions,

$$G_{ij}(t, t') = G_{ij}^0(t - t') + G_{ij}^1(t, t'), \tag{3.22}$$

where $G^1_{ij}(t, t')$ is not in principle a function of $(t - t')$. The equations of motion for the relevant Green's functions of zero and first order in the perturbation are given, respectively, by

$$i\frac{\partial}{\partial t}\langle\langle c_{i\sigma}(t); c^{+}_{n\sigma}(t')\rangle\rangle^0 = \delta(t - t')\delta_{in} + \sum_l t_{il}\langle\langle c_{l\sigma}(t); c^{+}_{n\sigma}(t')\rangle\rangle^0 + U\langle\langle n_{i-\sigma}(t)c_{i\sigma}(t); c^{+}_{n\sigma}(t')\rangle\rangle^0$$

and

$$i\frac{\partial}{\partial t}\langle\langle c_{i\sigma}(t); c^{+}_{n\sigma}(t')\rangle\rangle^1 = \sum_l t_{il}\langle\langle c_{l\sigma}(t); c^{+}_{n\sigma}(t')\rangle\rangle^1 + U\langle\langle n_{i-\sigma}(t)c_{i\sigma}(t); c^{+}_{n\sigma}(t')\rangle\rangle^1 - \sigma h e^{iq\cdot r_i}\langle\langle c_{i\sigma}(t); c^{+}_{n\sigma}(t')\rangle\rangle^0 e^{-i\omega t}.$$

Using a Hartree–Fock decoupling for the higher order Green's function and assuming that the system in the absence of the external magnetic field is paramagnetic, with no polarization, i.e., $\langle n_\sigma\rangle^0 = \langle n_{-\sigma}\rangle^0 = \langle n\rangle$, we get for the Fourier transformed, in time and space, zero order Green's function

$$\langle\langle c_{k\sigma}; c_{k'\sigma}\rangle\rangle^0_\omega = \frac{\delta_{kk'}}{\omega - E_k},$$

where $E_k = \epsilon_k + U\langle n\rangle$. The equation of motion for the first order Green's function in the Hartree–Fock approximation is,

$$i\frac{\partial}{\partial t}\langle\langle c_{i\sigma}(t); c^{+}_{n\sigma}(t')\rangle\rangle^1 = \sum_l t_{il}\langle\langle c_{l\sigma}(t); c^{+}_{n\sigma}(t')\rangle\rangle^1 + U\langle n\rangle\langle\langle c_{i\sigma}(t); c^{+}_{n\sigma}(t')\rangle\rangle^1 \quad (3.23)$$

$$+ U\delta n_{i-\sigma}(t)\langle\langle c_{i\sigma}(t); c^{+}_{n\sigma}(t')\rangle\rangle^0 - \sigma h e^{-iq\cdot r_i}e^{-i\omega t}\langle\langle c_{i\sigma}(t); c^{+}_{n\sigma}(t')\rangle\rangle^0. \quad (3.24)$$

We assume the magnetic polarization follows the spatial and temporal dependences of the field, i.e.,

$$\delta n_{i-\sigma}(t) = e^{-iq\cdot r_i}e^{-i\omega t}\delta n_{-\sigma}.$$

Fourier transforming equation (3.23) in lattice space $(\times\sum_i e^{ik\cdot r_i}\sum_n e^{-ik'\cdot r_n})$ yields

$$i\frac{\partial}{\partial t}\langle\langle c_{k\sigma}(t); c^{+}_{k'\sigma}(t')\rangle\rangle^1 = (\epsilon_k + U\langle n\rangle)\langle\langle c_{k\sigma}(t); c^{+}_{k'\sigma}(t')\rangle\rangle^1 + U\delta n_{-\sigma}e^{-i\omega t}\left\langle\left\langle c_{k-q,\sigma}(t); c^{+}_{k'\sigma}(t')\right\rangle\right\rangle^0 - \sigma h e^{-i\omega t}\left\langle\left\langle c_{k-q,\sigma}(t); c^{+}_{k'\sigma}(t')\right\rangle\right\rangle^0.$$

Next we Fourier transform in time, but only in the variable t, i.e., we write $c_{k\sigma}(t) = \int d\omega' e^{-i\omega' t}c_{k\sigma}(\omega')$, to obtain

$$\omega'\langle\langle c_{k\sigma}(\omega'); c_{k'\sigma}^{+}(t')\rangle\rangle_{\omega}^{1} = E_k\langle\langle c_{k\sigma}(\omega'); c_{k'\sigma}^{+}(t')\rangle\rangle_{\omega}^{1}$$
$$+ U\delta n_{-\sigma}\Big\langle\Big\langle c_{k-q,\sigma}(\omega + \omega'); c_{k',\sigma}^{+}(t')\Big\rangle\Big\rangle_{\omega}^{0}$$
$$- \sigma h\Big\langle\Big\langle c_{k-q,\sigma}(\omega + \omega'); c_{k'\sigma}^{+}(t')\Big\rangle\Big\rangle_{\omega}^{0}.$$

Then,

$$(\omega' - E_k)\langle\langle c_{k\sigma}(\omega'); c_{k'\sigma}^{+}(t')\rangle\rangle_{\omega}^{1} = [U\delta n_{-\sigma} - \sigma h]\Big\langle\Big\langle c_{k-q,\sigma}(\omega + \omega'); c_{k'\sigma}^{+}(t')\Big\rangle\Big\rangle_{\omega}^{0}.$$

Notice that the first order Green's function is expressed in terms of a zero order Green's function that depends only on $(t - t')$. Consequently, t' is arbitrary and can be taken as $t' = 0$. Consequently,

$$\Big\langle\Big\langle c_{k-q,\sigma}(\omega + \omega'); c_{k'\sigma}^{+}(t' = 0)\Big\rangle\Big\rangle_{\omega}^{0} = \frac{1}{\omega + \omega' - E_{k-q}}$$

and we get,

$$\langle\langle c_{k\sigma}(\omega'); c_{k'\sigma}^{+}(0)\rangle\rangle_{\omega}^{1} = \frac{U\delta n_{-\sigma} - \sigma h}{(\omega' - E_k)(\omega + \omega' - E_{k-q})}.$$

Since there is no more time dependence in this equation, we can use the fluctuation–dissipation theorem to calculate the correlation function $\delta n_{-\sigma}$. We find,

$$\delta n_\sigma = (U\delta n_{-\sigma} - \sigma h)\sum_{k,\omega'} F_\omega\left\{\frac{1}{(\omega' - E_k)(\omega + \omega' - E_{k-q})}\right\}.$$

Defining

$$\begin{aligned}\chi_0(q, \omega) &= \sum_{k,\omega'} F_\omega\left\{\frac{-1}{(\omega' - E_k)(\omega + \omega' - E_{k-q})}\right\}\\ &= \sum_k \frac{f(E_k) - f(E_{k+q})}{E_{k+q} - E_k - \omega}\end{aligned} \tag{3.25}$$

and $\delta m = \delta n_\uparrow - \delta n_\downarrow$, we finally obtain,

$$\delta m(q, \omega) = \frac{\chi_0(q, \omega)}{1 - U\chi_0(q, \omega)}h. \tag{3.26}$$

The dynamical susceptibility

For a free electron gas, with dispersion $\epsilon_k = \hbar^2k^2/2m$ the dynamical susceptibility $\chi_0(q, \omega)$ can be obtained [4]. We consider the case of three dimensions and zero temperature. A straightforward calculation yields for $q \ll k_F$,

$$\chi_0(q, \omega) \approx \rho(\epsilon_F)\left(1 + \frac{\omega}{2v_Fq}\ln\frac{\omega - v_Fq}{\omega + v_Fq}\right), \tag{3.27}$$

where $v_F = k_F/m$ is the Fermi velocity. More generally for $q \ll k_f$ and $\omega \ll v_F q$ we get,

$$\chi_0(q, \omega) \approx \rho(\epsilon_F)\left(1 + i\frac{\pi\omega}{2v_F q} - 12\frac{q^2}{k_F^2}\right), \tag{3.28}$$

such that the total susceptibility

$$\begin{aligned}\chi(q, \omega) = \frac{\chi_0(q, \omega)}{1 - U\chi_0(q, \omega)} &\approx \frac{\rho(\epsilon_F)}{1 - U\rho(\epsilon_F) - i\rho(\epsilon_F)\frac{\pi\omega}{2v_F q} + 12\rho(\epsilon_F)\frac{q^2}{k_F^2}} \\ &\approx \frac{1}{U_c - U - i\frac{\pi\omega}{2v_F q} + 12\frac{q^2}{k_F^2}},\end{aligned} \tag{3.29}$$

where $U_c = 1/\rho(\epsilon_F)$.

The imaginary part of $\chi(q, \omega)$ has a peak at

$$\omega_p = \frac{2v_F q}{\pi}\left(U_c - U + \frac{12q^2}{k_F^2}\right).$$

This peak characterizes the excitations of the paramagnetic state of a nearly ferromagnetic system, i.e., on the verge of satisfying the Stoner criterion. These excitations are known as *paramagnons* and differently from the spin-wave excitations of the ferromagnetic phase they are overdamped, since the width of the maximum of the imaginary part of $\chi(q, \omega)$ in momentum space is of the order, or larger than the value of the maximum itself. Physically this means that the paramagnons decay in time with a short lifetime.

Charge ordering instability

Let us consider the Hubbard Hamiltonian under the action of an external potential v that modulates the charge occupancy of the sites in the lattice,

$$H = \sum_{i,j,\sigma} t_{ij} c_{i\sigma}^{+} c_{j\sigma} + U\sum_i n_{i\uparrow}n_{i\downarrow} - ve^{-i\omega t}\sum_i e^{iq\cdot r_i} n_i,$$

where $n_i = n_{i\uparrow} + n_{i\downarrow}$. We want to calculate the charge density at site i. The relevant Green's function for this problem is $\langle\langle c_i(t); c_n^{+}(t')\rangle\rangle$ that obeys the following equation of motion,

$$\begin{aligned} i\frac{\partial}{\partial t}\langle\langle c_{i\sigma}(t); c_{n\sigma}^{+}(t')\rangle\rangle = \delta_{in} &+ \sum_l t_{il}\langle\langle c_{l\sigma}(t); c_{n\sigma}^{+}(t')\rangle\rangle \\ &+ U\langle\langle n_{i-\sigma}(t)c_{i\sigma}(t); c_{n\sigma}^{+}(t')\rangle\rangle \\ &- ve^{iq\cdot r_i}e^{-i\omega t}\langle\langle c_{i\sigma}(t); c_{n\sigma}^{+}(t')\rangle\rangle. \end{aligned}$$

In order to proceed we need to deal with the Green's function generated by the many-body interaction. This will be done by decoupling this Green's function in the following way,

$$U\langle\langle n_{i-\sigma}(t)c_{i\sigma}(t);\ c^{+}_{n\sigma}(t')\rangle\rangle^{0} \approx U\langle n\rangle\langle\langle c_{i\sigma}(t);\ c^{+}_{n\sigma}(t')\rangle\rangle^{1} + U\delta n_i\langle\langle c_{i\sigma}(t);\ c^{+}_{n\sigma}(t')\rangle\rangle^{0}$$

that corresponds to a mean-field or Hartree–Fock approximation. Now, we can gather terms of the same order in perturbation theory. Because in the absence of the external potential the system is uniform, the Fourier transformed, in space and time, zero order Green's function is given by,

$$\langle\langle c_k;\ c_{k'}\rangle\rangle^{0}_{\omega} = \frac{\delta_{kk'}}{\omega - E_k},$$

where $E_k = \epsilon_k + U\langle n\rangle$. The equation of motion for the first order Green's function in the RPA approximation is,

$$i\frac{\partial}{\partial t}\langle\langle c_{i\sigma}(t);\ c^{+}_{n\sigma}(t')\rangle\rangle^{1} = \sum_l t_{il}\langle\langle c_{l\sigma}(t);\ c^{+}_{n\sigma}(t')\rangle\rangle^{1} + U\langle n\rangle\langle\langle c_{i\sigma}(t);\ c^{+}_{n\sigma}(t')\rangle\rangle^{1} + U\delta n_i(t)\langle\langle c_{i\sigma}(t);\ c^{+}_{n\sigma}(t')\rangle\rangle^{0} - ve^{-iq\cdot r_i}e^{-i\omega t}\langle\langle c_{i\sigma}(t);\ c^{+}_{n\sigma}(t')\rangle\rangle^{0}.$$

The external potential is sufficiently well-behaved, such that,

$$\delta n_i(t) = e^{-iq\cdot r_i}e^{-i\omega t}\delta n.$$

Fourier transforming in lattice space ($\times\sum_i e^{ik\cdot r_i}\sum e^{-ik'\cdot r_n}$) yields

$$i\frac{\partial}{\partial t}\langle\langle c_{k\sigma}(t);\ c^{+}_{k'\sigma}(t')\rangle\rangle^{1} = (\epsilon_k + U\langle n\rangle)\langle\langle c_{k\sigma}(t);\ c^{+}_{k'\sigma}(t')\rangle\rangle^{1} + U\delta n\Big\langle\Big\langle c_{k-q,\sigma}(t);\ c^{+}_{k'\sigma}(t')\Big\rangle\Big\rangle^{0} - ve^{-i\omega t}\Big\langle\Big\langle c_{k-q,\sigma}(t);\ c^{+}_{k'\sigma}(t')\Big\rangle\Big\rangle^{0}.$$

Next we Fourier transform only in the variable t, i.e., we write $c_{k\sigma}(t) = \int d\omega' e^{-i\omega' t}c_{k\sigma}(\omega')$, to obtain

$$\omega'\langle\langle c_{k\sigma}(\omega');\ c^{+}_{k'\sigma}(t')\rangle\rangle^{1}_{\omega'} = E_k\langle\langle c_{k\sigma}(\omega');\ c^{+}_{k'\sigma}(t')\rangle\rangle^{1}_{\omega'} + U\delta n\Big\langle\Big\langle c_{k-q,\sigma}(\omega + \omega');\ c^{+}_{k'\sigma}(t')\Big\rangle\Big\rangle^{0}_{\omega'} - v\Big\langle\Big\langle c_{k-q,\sigma}(\omega + \omega');\ c^{+}_{k'\sigma}(t')\Big\rangle\Big\rangle^{0}_{\omega'}.$$

Then,

$$(\omega' - E_k)\langle\langle c_{k\sigma}(\omega'); c^{+}_{k'\sigma}(t')\rangle\rangle^1 = [U\delta n - v] \times \left\langle\left\langle c_{k-q,\sigma}(\omega + \omega'); c^{+}_{k'\sigma}(t')\right\rangle\right\rangle^0.$$

Notice that the first order Green's function is expressed in terms of a zero order Green's function that depends only in $(t - t')$. Consequently, t' is arbitrary and can be taken as $t' = 0$. We get

$$\left\langle\left\langle c_{k-q,\sigma}(\omega + \omega'); c^{+}_{k'\sigma}(t' = 0)\right\rangle\right\rangle^0 = \frac{1}{\omega + \omega' - E_{k-q}}$$

and,

$$\langle\langle c_{k\sigma}(\omega'); c^{+}_{k'\sigma}(0)\rangle\rangle^1 = \frac{U\delta n - v}{(\omega' - E_k)(\omega + \omega' - E_{k-q})}.$$

Since there is no more time dependence in the equation above, we can use the fluctuation dissipation theorem to calculate the correlation function δn. We get,

$$\delta n = (U\delta n - v)\chi_0(q, \omega),$$

where, as in the magnetic case, we have

$$\chi_0(q, \omega) = -\sum_{k,\omega'} F_\omega \left\{\frac{1}{(\omega' - E_k)(\omega + \omega' - E_{k-q})}\right\}.$$

We finally obtain

$$\begin{aligned}\delta n(q, \omega) &= \frac{\chi_0(q, \omega)}{1 + U\chi_0(q, \omega)} v \\ &\equiv \chi_c(q, \omega) v.\end{aligned} \tag{3.30}$$

For positive U there is no charge instability, since $\chi_c(q, \omega)$ does not diverge. However for negative U there may be a charge instability to a phase with a modulated charge density. This *charge density wave* state (CDW), as we saw previously, will compete with a superconducting ground state.

It is interesting to consider the full behavior of the dynamical charge susceptibility. The case $q \to 0$ is not interesting since this corresponds to a homogeneous charge distribution. Then, let us consider the case $\chi_0(q, 0)$ has a maximum for a finite value of $q = Q$. Expanding the charge susceptibility near $q = Q$ and for small frequencies $\omega \ll v_F Q$, we get for the charge susceptibility

$$\chi_c(Q, \omega) \approx \frac{\chi_0(Q)}{1 - |U|\chi_0(Q) - |q - Q|^2 - i\omega}, \tag{3.31}$$

where all finite constants have been made equal unity. For $q = Q$, the static part of the charge susceptibility diverges for $1 - |U|\chi_0(Q)$ signaling a charge instability at a quantum phase transition with $U_c = -1/\chi(Q)$.

It is worth comparing the expansion of the dynamic susceptibilities for the ferromagnetic homogeneous case ($q = 0$) and that close to a finite Q instability, equations (3.29) and (3.31), respectively. The dispersion of the paramagnons is distinct in each case. If we define a *dynamic critical exponent z* through the dispersion of the paramagnons, $\omega \propto q^z$ at the quantum critical point $U = U_c$, we find $z = 3$ and $z = 2$ for the zero q and finite Q quantum phase transitions, respectively. Then these phase transitions belong to different *universality classes*.

Finally, notice that in the magnetic case with $U > 0$ the same expansion of the dynamical susceptibility appears close to a magnetic instability to a spin-density wave state characterized by a wavevector Q.

References

[1] Tyablikov S V 1967 *Methods in the Quantum Theory of Magnetism* (New York: Plenum Press)

[2] Troper A, da Silva X A, Guimarães A P and Gomes A A 1975 *J. Phys. F: Metal Phys.* **5** 160

[3] Ramires A and Continentino M A 2011 *J. Phys.: Condens. Matter* **23** 125701

[4] Nozieres P 1986 *Magnétisme et Localisation dans les Liquides de Fermi* (Cours Collége de France) unpublished

Chapter 4

Green's functions and disorder

4.1 Configuration averaged Green's function

The problems studied so far, concerned lattice translation invariant systems and have used this property extensively to solve some complicated equations of motion using lattice Fourier transformation. The Green's functions obtained were diagonal in k-space. When one considers disordered systems this property is lost due to fluctuations caused in the parameters of the materials by the disorder and k ceases to be a good quantum number.

Disorder may be caused by different reasons and may have different levels. It may be due to doping or substitution, by the introduction of defects and alloying, for example. Glasses are disordered materials, perhaps the most radical form of disorder. It is remarkable that long wavelength excitations like phonons can propagate in glasses and spin waves in amorphous ferromagnets. Electronic transport properties are strongly affected by disorder. It may give rise to new effects as localization of the electronic wave functions. This opens a whole new area in condensed matter physics that requires new tools to be investigated.

Here we concentrate on questions concerning the propagation of magnetic [1] or elastic excitations in systems with disorder in the various forms of interactions or one-body hopping terms An essential step in the approach to these problems, especially within the Green's function formalism is to perform a configurational average of the Green's function. This average over all possible configurations restores translation invariance in the averaged system. This procedure allows the concept of k-space to remain useful. The price to pay is that now the excitations are damped, i.e., the time dependent Green's function possesses a real part that describes the finite lifetime of these modes. Unfortunately, not all problems with disorder can be treated in this way. Instead of average quantities some problems require the calculation of their probability distributions, as in the case of localization.

Consider a disordered ferromagnetic material described by the Heisenberg Hamiltonian, but with exchange interactions that can fluctuate from site to site.

doi:10.1088/978-0-7503-3395-5ch4

The equation of motion for the transverse Green's function that describes spin-wave propagation is given by

$$i\frac{d}{dt}\left\langle\left\langle S_i^+(t);\, S_j^-(0)\right\rangle\right\rangle = -\,2\delta(t)\langle S_i^z\rangle\delta_{ij} + \sum_l J_{il}\left\langle\left\langle S_l^z S_i^+(t);\, S_j^-(0)\right\rangle\right\rangle$$
$$-\sum_l J_{il}\left\langle\left\langle S_i^z S_l^+(t);\, S_j^-(0)\right\rangle\right\rangle, \tag{4.1}$$

where the J_{ij} are random variables with a given probability distribution.

At this point we make a random phase approximation to decouple the higher order Green's function. We also assume that the magnetization is uniform throughout the system, i.e., $\langle S_i^z\rangle = \langle S^z\rangle = \sigma$. This is not a bad approximation at least to describe long wavelength spin waves. Defining,

$$G_{ij}^{+-}(t) = \left\langle\left\langle S_i^+(t);\, S_j^-(0)\right\rangle\right\rangle,$$

we obtain

$$i\frac{d}{dt}G_{ij}^{+-}(t) = -2\delta(t)\langle S^z\rangle\delta_{ij} + \langle S^z\rangle\sum_l J_{il}G_{ij}^{+-}(t) - \langle S^z\rangle\sum_l J_{il}G_{lj}^{+-}(t).$$

A formal solution of this equation can be obtained as,

$$G_{ij}^{+-}(t) = 2i\sigma\theta(t)\left\{e^{i\sigma t\hat{\Omega}}\right\}_{ij}, \tag{4.2}$$

where the matrix $\hat{\Omega}$ has elements given by

$$\Omega_{ij} = J_{ij} - \delta_{ij}\sum_l J_{il}.$$

It is useful to consider the Fourier transformations of this equation, i.e.,

$$G_{kk'}^{+-}(t) = (1/N)\sum_i\sum_j e^{ik\cdot r_i}e^{ik'\cdot r_j}G_{ij}^{+-}(t),$$

such that

$$G_{kk'}^{+-}(t) = 2i\sigma\theta(t)\rho(k-k')\langle k|e^{-i\sigma t\sum_{k_1k_2}\Omega_{k_1k_2}|k_1\rangle\langle k_2|}|k'\rangle \tag{4.3}$$

where

$$\rho(k-k') = (1/N)\sum_i e^{-i(k-k')\cdot r_i}$$

and

$$\Omega_{k_1k_2} = (1/N)\sum_i e^{-i(k_1-k_2)\cdot r_i}\sum_j J_{ij}(1 - e^{-ik_2\cdot(r_i-r_j)}).$$

For a translation invariant, ordered system, $\rho(k - k') = \delta_{kk'}$ and the spin-wave propagator is diagonal in k-space. Furthermore, $\Omega_{k_1k_2} = \omega_{k_1}\delta_{k_1k_2}$ and the spectral density, i.e., the imaginary part of the time Fourier transform of the Green's function is given by $(1/\pi)ImG^{+-}_{kk'}(\omega) = \delta(\omega - \omega_{k_1})$. In this case the spin-waves are sharp excitations with a well-defined momentum.

In equation (4.3), we have chosen to write the non-diagonal Green's function using a ket representation for the sake of clarity. These kets span an abstract Hilbert space and are orthonormalized. The Green's function in equation (4.3) corresponds to a given configuration of the exchange interactions J_{ij} and is non-diagonal in k-space.

We consider that the random exchange interactions are distributed according to a Gaussian,

$$J_{ij}(r_i, r_j) = J_0(r_i - r_j) + j_{ij}(r_i, r_j) \tag{4.4}$$

$$\begin{cases} \text{with } \langle j_{ij}\rangle_{AV} & = 0 \\ \text{and } \langle j_{fg}j_{hk}\rangle_{AV} & = j^2(\delta_{fh} \times \delta_{gk} + \delta_{fk} \times \delta_{gh}) \end{cases} \tag{4.5}$$

where $\langle\cdots\rangle_{AV}$ means a configurational average over all possible realizations of the distribution of exchange interactions.

Next step is to perform the configurational average of the Green's function, equation (4.3) to restore translation invariance. This is carried out in the following way,

$$\langle G^{+-}_{kk'}(t)\rangle_{AV} = 2i\sigma\theta(t)\langle\rho(k - k')\rangle_{AV}\langle\langle\langle k|e^{-i\sigma t\sum_{k_1k_2}\Omega_{k_1k_2}|k_1\rangle\langle k_2|}|k'\rangle\rangle\rangle_{AV}.$$

Since, $\langle\rho(k - k')\rangle_{AV} = \delta_{kk'}$, this guarantees the restoration of translation invariance in the system and the average Green's function becomes diagonal in k-space.

$$\langle G^{+-}_{kk'}(t)\rangle_{AV} = 2i\sigma\theta(t)\langle\langle\langle k|e^{-i\sigma t\sum_{k_1k_2}\Omega_{k_1k_2}|k_1\rangle\langle k_2|}|k\rangle\rangle\rangle_{AV}.$$

Defining a diagonal operator associated with the average spin-wave energy,

$$\Omega_0 = \sum_{k_1k_2}\langle\Omega_{k_1k_2}\rangle_{AV}|k_1\rangle\langle k_2| = \sum_k \omega(k)|k\rangle\langle k|,$$

where

$$\omega(\mathbf{k}) = \langle\Omega_{k_1k_2}\rangle_{AV} = J_0\sum_{i=1,d}(1 - \cos k_i a) \tag{4.6}$$

with a an average interatomic distance between the magnetic moments in a hypercubic lattice of dimension d. In the limit $k \to 0$, $\omega(k) = Dk^2$ where D is the spin wave stiffness. The mode at $k = 0$ with energy zero is the Goldstone mode that restores the original rotational symmetry of the Heisenberg Hamiltonian broken by the appearance of a ferromagnetic state with a preferred magnetization direction along the z-axis. This Goldstone mode as we show below is *unaffected by disorder*.

Using the identity

$$e^{-i\sigma t\hat{\Omega}} \equiv e^{-it\sigma\hat{\Omega}_0}\mathcal{T}e^{-i\sigma\int_0^t d\tau\hat{\Omega}_1(\tau)},$$

where

$$\hat{\Omega}_1(\tau) = e^{-i\sigma\hat{\Omega}_0\tau}(\hat{\Omega} - \hat{\Omega}_0)e^{i\sigma\hat{\Omega}_0\tau}$$

and $\mathcal{T}$ the time ordering operator we met before in chapter 1. We can write the average Green's function as

$$\langle G_{kk'}^{+-}(t)\rangle_{AV} = 2i\sigma\theta(t)e^{-i\sigma\omega(k)t}\langle(\langle k|\mathcal{T}e^{-i\sigma\int_0^t d\tau\hat{\Omega}_1(\tau)}|k\rangle)\rangle_{AV}.$$

4.2 Spin wave propagation in disordered media as a random frequency modulation problem

In order to calculate the configurational average of the time ordered exponential, i.e.,

$$\langle(\langle k|\mathcal{T}e^{-i\sigma\int_0^t d\tau\hat{\Omega}_1(\tau)}|k\rangle)\rangle_{AV},$$

we consider a *cumulant expansion* and keep terms only to second order, consistent with the Gaussian distribution of exchange interactions [2], equation (4.5).

Let us define the average of an operator A that we represent as ((A)) by

$$((A)) = \langle\langle k|A|k\rangle\rangle_{AV}. \tag{4.7}$$

Since $((1)) = 1$ and $((aA + bB)) = a((A)) = b((B))$, this a proper averaging procedure. Then we can rewrite the configurational averaged Green's function as,

$$\langle G_{kk'}^{+-}(t)\rangle_{AV} = 2i\sigma\theta(t)e^{-i\sigma\omega(k)t}\left(\left(Te^{-i\sigma\int_0^t d\tau\hat{\Omega}_1(\tau)}\right)\right). \tag{4.8}$$

The Gaussian approximation corresponds to consider terms up to second order in the cumulant expansion [2]. We get,

$$\left(\left(Te^{-i\sigma\int_0^t d\tau\hat{\Omega}_1(\tau)}\right)\right) \approx e^{-i\sigma\int_0^t d\tau((\hat{\Omega}_1(\tau)))_c - \frac{1}{2}\sigma^2\int_0^t d\tau_1\int_0^t d\tau_2((T[\hat{\Omega}_1(\tau_1)\hat{\Omega}_1(\tau_2)]))_c}. \tag{4.9}$$

The first cumulant average is defined by,

$$((\hat{\Omega}_1(\tau_1)))_c = (([\hat{\Omega}(\tau_1) - \hat{\Omega}_0(\tau_1)])) = 0 \tag{4.10}$$

since $(([\hat{\Omega}(\tau_1)])) = \langle\langle k|\hat{\Omega}_1(\tau)|k\rangle\rangle_{AV} = \hat{\Omega}_0 = \omega(k)$. For the second order cumulant average we get,

$$\begin{aligned}
\langle\langle T[\hat{\Omega}_1(\tau_1)\hat{\Omega}_1(\tau_2)]\rangle\rangle_c &= \langle\langle T[\hat{\Omega}(\tau_1) - \hat{\Omega}_0(\tau_1)][\hat{\Omega}(\tau_2) - \hat{\Omega}_0(\tau_2)]\rangle\rangle_c \\
&= \langle\langle T[\hat{\Omega}(\tau_1) - \hat{\Omega}_0(\tau_1)][\hat{\Omega}(\tau_2) - \hat{\Omega}_0(\tau_2)]\rangle\rangle \\
&- \langle\langle[\hat{\Omega}(\tau_1) - \hat{\Omega}_0(\tau_1)]\rangle\rangle\langle\langle[\hat{\Omega}(\tau_2) - \hat{\Omega}_0(\tau_2)]\rangle\rangle \\
&= \langle\langle T[\hat{\Omega}(\tau_1) - \hat{\Omega}_0(\iota_1)][\hat{\Omega}(\iota_2) - \hat{\Omega}_0(\tau_2)]\rangle\rangle \\
&= \langle\langle T[\hat{\Omega}_1(\tau_1)\hat{\Omega}_1(\tau_2)]\rangle\rangle,
\end{aligned}$$

where we used equations (4.7) and (4.10).

Then, at the Gaussian level of approximation, we get

$$\begin{aligned}
\left\langle\left\langle Te^{-i\sigma\int_0^t d\tau\hat{\Omega}_1(\tau)}\right\rangle\right\rangle &= e^{-\frac{1}{2}\sigma^2\int_0^t d\tau_1\int_0^t d\tau_2\langle\langle T[\hat{\Omega}_1(\tau_1)\hat{\Omega}_1(\tau_2)]\rangle\rangle} \\
&= e^{-\gamma(t)},
\end{aligned}$$

where

$$\gamma(t) = \sigma^2\int_0^t d\tau_1\int_0^{\tau_1} d\tau_2\langle\langle\hat{\Omega}_1(\tau_1)\hat{\Omega}_1(\tau_2)\rangle\rangle.$$

Since the correlation function,

$$F(\tau_1, \tau_2) = \langle\langle\hat{\Omega}_1(\tau_1)\hat{\Omega}_1(\tau_2)\rangle\rangle$$

depends only on the difference $\tau = \tau_1 - \tau_2$, we can write,

$$\gamma(t) = \sigma^2\int_0^t d\tau(t - \tau)F(\tau), \tag{4.11}$$

and for the average Green's function,

$$\langle G_k^{+-}(t)\rangle_{AV} = 2i\sigma\theta(t)e^{-i\sigma\omega(k)t}e^{-\gamma(t)}.$$

The quantity $\gamma(t)$ contains the results of perturbation theory, to second order in the strength of disorder, for the spin-wave energy and for the transition probabilities. We will neglect the former.

It turns out to be interesting to look at the problem of spin-wave propagation in a random media from the point of view of the theory of stochastic frequency modulation [2]. In this framework, disorder modulates the spin-wave propagation, represented by the term $\exp(-i\sigma\omega(k)t)$, through the Gaussian random process $\Omega_1(\tau)$.

The correlation function $F(\tau)$ of the random process $\hat{\Omega}_1(\tau)$ can be characterized by two parameters,

- a k-dependent amplitude,

$$\Delta^2(k) = F(\tau = 0),$$

- and a k-dependent correlation time given by,

$$\tau_c(k) = \frac{\int_0^\infty ((\hat{\Omega}_1(\tau)\hat{\Omega}_1(0)))d\tau}{\Delta^2} = \frac{\int_0^\infty F(\tau)d\tau}{\Delta^2}.$$

These quantities can be calculated using the definition of the averaging procedure $((\ldots))$. We obtain

$$\Delta^2(k) = \sum_{k'} \langle \Omega_{kk'}\Omega_{k'k} \rangle_{AV} - (\omega(k))^2 \tag{4.12}$$

and

$$\begin{aligned} \tau_c(k) &= \frac{2\pi \sum_{k'} \delta(\omega(k) - \omega(k'))[\langle \Omega_{kk'}\Omega_{k'k} \rangle_{AV} - (\omega(k))^2]}{\Delta^2(k)} \\ &\approx \frac{2\pi[\langle \Omega_{kk}\Omega_{kk} \rangle_{AV} - (\omega(k))^2] \sum_{k'} \delta(\omega(k) - \omega(k'))}{\Delta^2(k)}, \end{aligned} \tag{4.13}$$

where the remaining average $\langle \cdots \rangle_{AV}$ is performed using the distribution of exchange interactions, equation (4.5).

The correlation time $\tau_c(k)$ is a measure of the speed of modulation, such that $F(\tau) = ((\hat{\Omega}_1(\tau)\hat{\Omega}_1(0))) \approx 0$ for $t \gg \tau_c$.

Notice that we can distinguish two regimes of time where the function $\gamma(t)$ behaves differently.

- $t \ll \tau_c$

  In this region the correlation function $F(\tau)$ is approximately constant and equal to $F(\tau = 0)$. The time integral in equation (4.11) can be easily performed and we obtain that in this regime $\gamma(t)$ behaves like

$$\gamma(t) = \frac{1}{2}\sigma^2\Delta^2(k)t^2,$$

  with $\Delta^2(k)$ given by equation (4.12). Then, in this *short time* regime the spin-wave Green's function is a Gaussian given by

$$\langle G_k^{+-}(t) \rangle_{AV} = 2i\sigma\theta(t)e^{-i\sigma\omega(k)t - \frac{1}{2}\sigma^2\Delta^2(k)t^2}, \tag{4.14}$$

  and the corresponding spectral density is also a Gaussian.

- $t \gg \tau_c$

  Since $F(\tau) \approx 0$ for $\tau \gg \tau_c$, we can push the limit of the integral in equation (4.11) to infinity and obtain

$$\gamma(t) = \sigma^2|t|\int_0^\infty F(\tau)d\tau + \text{constant}.$$

The first term in $\gamma(t)$ contains essentially the results of perturbation theory. The integral in $F(\tau)$ has an imaginary part that gives the correction to the spin-wave energy to second order and that we neglect. Its real part yields the relaxation of the spin-waves and is given by,

$$\gamma_R(t) = \sigma^2|t|\Delta^2(k)\tau_c(k)$$

with $\Delta(k)$ and $\tau_c(k)$ defined in equations (4.12) and (4.13). Then in the long time regime, the Green's function is given by

$$\langle G_k^{+-}(t)\rangle_{AV} = 2i\sigma\theta(t)e^{-i\sigma\omega(k)t-\sigma^2\Gamma(k)t}, \tag{4.15}$$

and the corresponding spectral density is a Lorentzian. The *damping* term $\Gamma(k) = \Delta^2(k)\tau_c(k)$.

We wish to decide which regime gives the correct description of the spin-wave propagation in the presence of disorder, in particular in the hydrodynamic limit $k \to 0$. For this purpose, we have to pursue our analogy with the random frequency modulation problem. In this approach we can distinguish two situations depending on the relative magnitudes of Δ and τ_c.

- Slow modulation $\Delta\tau_c \gg 1$.
  In this case the correlation time $\tau_c(k)$ is large compared with $1/\Delta(k)$. The Green's function is represented by the Gaussian propagator and only for very large values of time, exponential behavior sets in. The spectral density is a Gaussian with linewidth $\Delta(k)$.
- Fast modulation $\Delta\tau_c \ll 1$.
  The correlation time $\tau_c(k)$ is small compared with $1/\Delta(k)$. The modulation due to the random process $\hat{\Omega}_1(\tau)$ is hardly effective and averages out. The spin-wave propagator becomes sharp and centered around the spin-wave energy. The spectral density is a Lorentzian with half-width $\sigma^2\Gamma(k) = \sigma^2\Delta^2(k)\tau_c(k)$. Since the spectral density becomes narrower in this regime, the condition $\Delta\tau_c \ll 1$ is identified as a *narrowing condition.*

The configuration averages in equations (4.12) and (4.13) can be calculated using the distribution of exchange interactions given in equation (4.5). Since we are interested in the hydrodynamic limit $k \to 0$, we get for the parameters characterizing the random process in this limit (see section 4.5.1):

- The amplitude of modulation

$$\Delta^2(k) = \sum_{k'}\langle\Omega_{kk'}\Omega_{k'k}\rangle_{AV} = 3j^2\sum_{j}(1 - \cos k\cdot(r_i - r_j)).$$

For a hypercubic lattice in the limit $k \to 0$, we find

$$\Delta^2(k) = 3j^2a^2k^2,$$

where a is the interatomic distance in the *pure*, ordered system.

- Correlation time

$$\tau_c(k) \approx \frac{2\pi[\langle\Omega_{kk}\Omega_{kk}\rangle_{AV} - (\omega(k))^2]\rho_d(k)}{\Delta^2(k)},$$

where $\rho_d(k)$ is the density of states. For a hypercubic lattice in the limit $k \to 0$ we get,

$$\tau_c(k) = \frac{1}{3\pi J_0}(ak)^3.$$

Finally, we can express the k-dependent *narrowing condition* as,

$$\Delta(k)\tau_c(k) = \frac{j}{\sqrt{3}\,\pi J_0}(ak)^4 \ll 1. \tag{4.16}$$

Notice that in the hydrodynamic limit this condition of fast modulation is always realized. The Green's function in this limit is given by equation (4.15), where $\Gamma(k) = \Delta^2(k)\tau_c(k) = (j^2/2\pi^2 J_0)(ak)^5$. The spectral density is a Lorentzian,

$$Im\langle G_k^{+-}(\omega)\rangle_{AV} = \frac{1}{\pi}\frac{\Gamma(k)}{(\omega - \omega(k))^2 - \Gamma^2(k)}.$$

Since the damping $\Gamma(k) = \Delta^2(k)\tau_c(k)$, the fast modulation condition given by $\Delta(k)\ \tau_c(k) \ll 1$ implies that $\Gamma(k) \ll \Delta^2(k)$ so that the spectral density is narrowed in the hydrodynamic limit. Then the fast modulation condition is indeed a *narrowing condition*. It is a consequence here of taking the hydrodynamic limit $k \to 0$ and as such it can be identified as a *hydrodynamic narrowing*.

This hydrodynamic narrowing condition may be responsible for low energy excitations being well-defined in disordered systems, such as magnons in amorphous ferromagnets and phonons in glasses.

The approach developed above provides a theory of the line-shape of spin-waves as measured, for example, by inelastic neutron scattering since the neutron cross section is related to the spectral function of the Green's function. Here we have obtained results for long wavelength magnons which are described by Lorentzian line shapes as we have shown.

Another interesting point related to hydrodynamic narrowing is that a kind of irreversibility is introduced when taking the limit $k \to 0$. This can be easily seen if we consider the damping part of the Green's function when an arbitrary reference time t_0 is chosen. In the Gaussian case this is

$$\exp[-(1/2)\sigma^2\Delta^2(t - t_0)^2],$$

and in the case the hydrodynamic condition is satisfied we have

$$\exp[-\sigma^2\Gamma(t - t_0)].$$

While in the former there is a persistence or *remembrance* of the initial conditions, in the latter case this introduces just a multiplying factor that does not affect the structure of the Green's function.

4.3 The infinite range ferromagnet

There are few problems with disorder in statistical mechanics that have an exact solution. Even when these models are simple and idealized, their solutions still can provide deep insight. One of these models is that of a ferromagnet described by the Heisenberg Hamiltonian with random infinite range interactions [3]. Let us start considering the ordered system where the interactions J_{ij} between any pair of moments are equal [4]. The Hamiltonian is given by,

$$H = -\frac{J}{N}\sum_{ij}\mathbf{S}_i \cdot \mathbf{S}_j,$$

where N is the number of ions and the normalization is necessary to yield a meaningful result. As in the previous section a formal solution of the transverse Green's function can be written as

$$G_{ij}^{+-}(t) = -2i\sigma\theta(t)[e^{i\sigma t(\mathbf{\Omega}_1+\mathbf{\Omega}_b)}]_{ij},$$

where the matrices

$$(\mathbf{\Omega}_1)_{ij} = \frac{J}{N}$$
$$(\mathbf{\Omega}_2)_{ij} = -\ J\delta_{ij}.$$

The matrix $\mathbf{\Omega}_1$ has all elements equal to J/N. Since the diagonal matrix $\mathbf{\Omega}_2$ commutes with $\mathbf{\Omega}_1$, we can write,

$$G_{ij}^{+-}(t) = -2i\sigma\theta(t)e^{-i\sigma tJ}[e^{i\sigma t\mathbf{\Omega}_1}]_{ij},$$

where $[\mathbf{A}]_{ij}$ means the element (ij) of the matrix $\mathbf{A}$. Notice that we can write,

$$[e^{i\sigma t\mathbf{\Omega}_1}]_{ij} = \sum_l a_{li}a_{lj}e^{i\sigma t\lambda_l},$$

where a_{li} is the ith component of the eigenvector of the eigenvalue λ_l of the matrix $\mathbf{\Omega}_1$. In order to calculate the magnetization, we only need the diagonal Green's function $G_{ii}^{+-}(t)$ and since the eigenvectors are orthonormalized, we get

$$G_{ii}^{+-}(t) = -2i\sigma\theta(t)e^{-i\sigma tJ}\left(\sum_l e^{i\sigma t\lambda_l}\right).$$

The normalized density of eigenvalues of the matrix $\mathbf{\Omega}_1$ ($[\mathbf{\Omega}_1]_{ij} = J/N$) is given by,

$$\rho(\lambda) = \frac{1}{N}\delta(\lambda - J) + \frac{N-1}{N}\delta(\lambda).$$

We can then write,

$$\sum_l e^{i\sigma t\lambda_l} = \int d\lambda \rho(\lambda) e^{i\sigma t\lambda} = \frac{1}{N}e^{i\sigma tJ} + \frac{N-1}{N}$$

and

$$\begin{aligned} G_{ii}^{+-}(t) &= -2i\sigma\theta(t)e^{-i\sigma tJ}\left(\frac{1}{N}e^{i\sigma tJ} + \frac{N-1}{N}\right) \\ &= -2i\sigma\theta(t)\left(\frac{1}{N} + \frac{N-1}{N}e^{-i\sigma tJ}\right). \end{aligned}$$

A Fourier transformation in time yields,

$$G_{ii}^{+-}(\omega) - \frac{2}{2\pi}\left(\frac{1}{N}\frac{\sigma}{\omega + i\epsilon} + \frac{N-1}{N}\frac{\sigma}{\omega - \sigma J + i\epsilon}\right)$$

and

$$ImG_{ii}^{+-}(\omega) = \frac{1}{N}\delta(\omega/\sigma) + \frac{N-1}{N}\delta(\omega/\sigma - J). \tag{4.17}$$

The single, zero energy excitation is the Goldstone, symmetry restoring mode that is present even in the Heisenberg model with infinite range interactions. It represents a uniform rotation of the whole system and is a consequence of the rotation symmetry of the Hamiltonian.

The self-consistent equation for the magnetization is given by

$$\sigma = S - 2\int d\omega \frac{ImG_{ii}^{+-}(\omega)}{e^{\beta\omega} - 1}. \tag{4.18}$$

The diverging contribution due the Goldstone mode is of order O(1/N) and can be neglected in the hydrodynamic limit. In fact the correct way to deal with this contribution is to consider an infinitesimal magnetic field parallel to the magnetization, and then take the thermodynamic limit.

Equation (4.18) yields the following self-consistent equation for the magnetization

$$\sigma = S\tanh(\beta J\sigma/2). \tag{4.19}$$

This is the well-known mean-field equation for the magnetization of a spin-1/2 ferromagnet ($S = 1/2$), showing that the exact solution of the infinite range model is the mean-field result, as expected.

4.4 The infinite range random Heisenberg ferromagnet

The transverse Green's function can be written as before. The elements of the matrices $\mathbf{\Omega}_1$ and $\mathbf{\Omega}_2$ are given by,

$$(\mathbf{\Omega}_1)_{ij} = J_{ij}/N$$
$$(\mathbf{\Omega}_2)_{ij} = -\,\delta_{ij}(1/N)\sum_l J_{il},$$

where J_{ij} are infinite range random interactions. Each of the elements of the diagonal matrix $\mathbf{\Omega}_2$ is a sum of a very large number of contributions. By the central limit theorem these *local fields* should be distributed according to a Gaussian, with a very narrow distribution centered at the mean-value $\bar{\omega}_0$, independent of a particular site i. Since a constant diagonal matrix commutes with any matrix, we can write the average Green's function as,

$$\langle G_{ij}^{+-}(t)\rangle_{AV} = -2i\sigma\theta(t)e^{-i\sigma\bar{\omega}_0 t}\langle[e^{i\sigma t\mathbf{\Omega}_1}]_{ij}\rangle_{AV}. \tag{4.20}$$

The existence of a Goldstone mode in the Heisenberg ferromagnet is a consequence of the rotational invariance of the model and independent whether the interactions are random or not. Mathematically, this arises since the matrix $\mathbf{\Omega}_1 + \mathbf{\Omega}_2$ has zero determinant that guarantees that at least one eigenvalue is zero. When we decouple the equation of motion and the matrices $\mathbf{\Omega}_1$ and $\mathbf{\Omega}_2$, as in equation (4.20), this condition may be lost and needs to be reintroduced afterwards through a consistent choice of $\bar{\omega}_0$.

The configurational average will be carried out assuming that the random infinite range interactions J_{ij} are distributed according to a Gaussian,

$$P(J_{ij}) = \frac{1}{\sqrt{2\pi}\tilde{J}}e^{\frac{(J_{ij}-J_0/N)^2}{2\tilde{J}^2}},$$

where $\tilde{J} = J/\sqrt{N}$, and this is appropriately normalized to give consistent results for large N. In this case the distribution of the eigenvalues θ_λ of the random matrix $\mathbf{\Omega}_1$ is given by the semi-circular law [5],

$$\rho_0(\theta_\lambda) = \begin{cases} \dfrac{\sqrt{4J^2 - \theta_\lambda^2}}{2\pi J^2}, & \text{for } |\theta_\lambda| < 2J \\ 0, & \text{for } |\theta_\lambda| \geqslant 2J. \end{cases}$$

If $J_0 > J$ there is an extra eigenvalue and the density of states is [5],

$$\rho(\theta_\lambda) = \rho_0(\theta_\lambda) + \frac{1}{N}\delta(\theta_\lambda - J_E), \tag{4.21}$$

where

$$J_E = J_0 + \frac{J^2}{J_0}.$$

In order to obtain the configuration average in equation (4.20), we proceed as in the pure case and write,

$$\langle[e^{i\sigma t\mathbf{\Omega}_1}]_{ij}\rangle_{AV} = \sum_\lambda \langle a_{i\lambda}a_{\lambda j}\rangle_{AV}\langle e^{i\sigma t\theta_\lambda}\rangle_{AV},$$

where $a_{i\lambda}$ are the components of the eigenvector of the eigenvalue θ_λ. The separable nature of the average is due to the statistical independence of the elements J_{ij} of the matrix $\mathbf{\Omega}_1$. Since we are interested in the local Green's function, and the eigenvectors of $\mathbf{\Omega}_1$ are orthonormalized, we have

$$\begin{aligned}\langle G_{ii}^{+-}(t)\rangle_{AV} &= -\,2i\sigma\theta(t)e^{-i\sigma\bar{\omega}_0 t}\langle[e^{i\sigma t\mathbf{\Omega}_1}]_{ii}\rangle_{AV} \\ &= -\,2i\sigma\theta(t)e^{-i\sigma\bar{\omega}_0 t}(1/N)\sum_\lambda\langle e^{i\sigma t\theta_\lambda}\rangle_{AV}.\end{aligned}$$

For the last average, we have

$$(1/N)\sum_\lambda\langle e^{i\sigma t\theta_\lambda}\rangle_{AV} = \int d\theta_\lambda\rho(\theta_\lambda)e^{i\sigma t\theta_\lambda}.$$

Using equation (4.21), and performing the integrations, we finally obtain

$$\langle G_{ii}^{+-}(t)\rangle_{AV} = -2i\sigma\theta(t)\left\{2\frac{J_1(2J\sigma t)}{2J\sigma t}e^{-i\sigma\bar{\omega}_0 t} + \frac{1}{N}e^{-i\sigma(\bar{\omega}_0 - J_E)t}\right\}, \tag{4.22}$$

where $J_1(x)$ is the first order cylindrical Bessel function and the second term exists only for $J_0 > J$. The first term represents the oscillatory part of the magnetic excitations and their damping that for large times decreases as $t^{-3/2}$. Notice that this a small decay compared to an exponential damping. The second term describes a purely oscillatory excitation with energy, $(\bar{\omega}_0 - J_E)$. The spectral density, i.e., the imaginary part of the local Green's function is obtained Fourier transforming in time the equation above, we get

$$Im\langle G_{ii}^{+-}(\omega)\rangle_{AV}^0 = \begin{cases}\dfrac{\sqrt{4J^2 - (\omega/\sigma - \bar{\omega}_0)^2}}{2\pi J^2}, & \text{for } |\omega/\sigma - \bar{\omega}_0| \leqslant 2J \\ 0, & \text{for } |\omega/\sigma - \bar{\omega}_0| > 2J\end{cases}$$

and for $J_0 > J$,

$$Im\langle G_{ii}^{+-}(\omega)\rangle_{AV} = Im\langle G_{ii}^{+-}(\omega)\rangle_{AV}^0 + \frac{1}{N}\delta(\omega/\sigma - (\bar{\omega}_0 - J_E)). \tag{4.23}$$

The stability of the ferromagnetic phase requires that $\omega Im\langle G_{ii}^{+-}(\omega)\rangle_{AV} \geqslant 0$ (condition for the absorption of an external energy by the elementary excitations). This, together with the fact that the determinant of $\mathbf{\Omega}_1 + \mathbf{\Omega}_2$ vanishes imposes a constraint on the energy of the isolated mode. Then for $J_0 > J$, such that this mode exists, its energy must be zero and it can be identified with the Goldstone mode of the disordered ferromagnet, as shown in figure 4.1. Then we have,

$$\bar{\omega}_0 = J_0 + \frac{J^2}{J_0}. \tag{4.24}$$

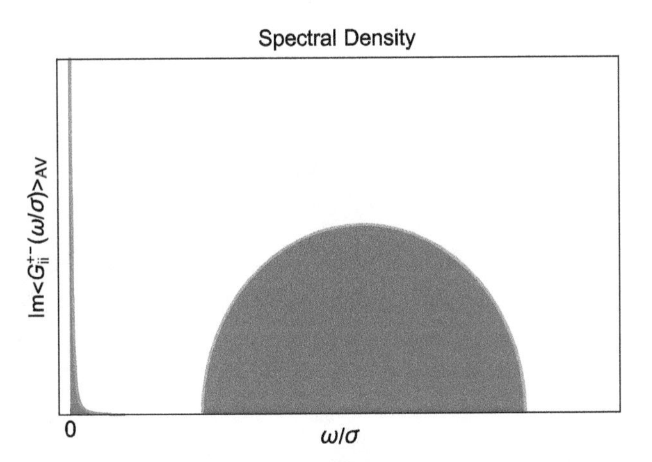

Figure 4.1. The spectral density for $J_0 > J$ (schematic) showing the isolated eigenvalue at zero energy.

From now on we focus in the ferromagnetic case $J_0 > J$. For $J_0 < J$ a finite density of states at zero frequency leads to an instability of the ferromagnetic phase and the ground state is a spin glass. Using the condition equation (4.24) and the result for the spectral density, we get the following result for the magnetization

$$\frac{S}{\sigma} = \frac{1}{2\pi J^2} \int_{\bar{\omega}_0 - 2J}^{\bar{\omega}_0 + 2J} d\omega \sqrt{4J^2 - (\omega - \bar{\omega}_0)^2} \coth(\beta\sigma\omega/2), \tag{4.25}$$

where $S = 1/2$. The Curie temperature can be easily obtained and is given by

$$k_B T_C = \frac{J_0}{4}$$

independent of the width of the distribution of exchange interactions. The magnetization curve of the disordered ferromagnet is shown in figure 4.2.

As we already pointed out, the method of averaging the single particle Green's function is not always appropriate to deal with disordered systems. A dramatic example is that of localization in electronic systems. In this problem instead of average quantities, it is the typical values that are measured in experiments. In this case the relevant information is contained in the probability distribution and all its moments and not in average values.

4.5 Appendix

4.5.1 Calculation of the amplitude and the correlation time

1. Average spin-wave energy

$$\langle \Omega_{kk'} \rangle_{AV} = \omega(k) = \delta_{kk'} J_0 \sum_j (1 - e^{ik\cdot(r_i - r_j)}).$$

For a hypercubic lattice in the hydrodynamic limit ($k \to 0$) we obtain

$$\omega(k) = J_0 a^2 k^2$$

where a is the average nearest neighbor distance.

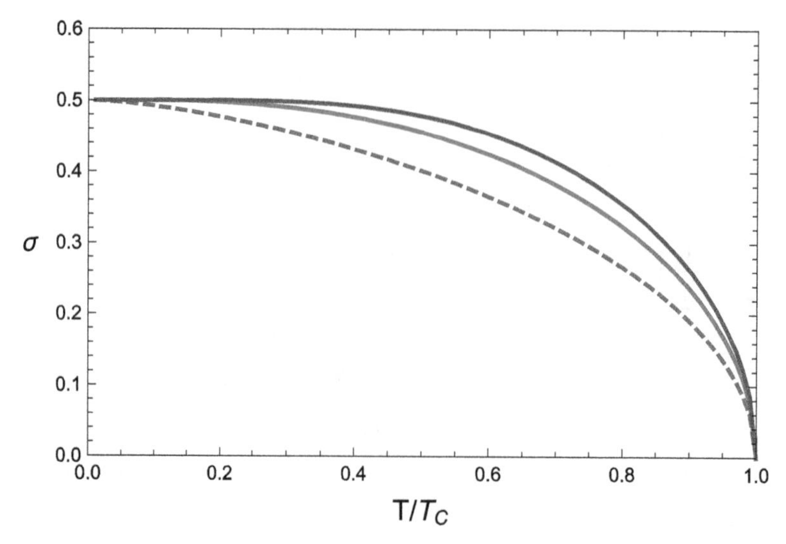

Figure 4.2. *Flattening* of the magnetization curve of the infinite range disordered $S = 1/2$ ferromagnet, by disorder. Purple curve is the mean-field result of the pure system, equation (4.19). Red is for $J/J_0 = 1/2$ and blue dashed for $J/J_0 = 1$, equation (4.25).

2. Second moment

$$\Delta^2(k) = \sum_{k'} \langle \Omega_{kk'}\Omega_{k'k}\rangle_{AV} = 3j^2 \sum_{j}(1 - \cos k \cdot (r_i - r_j))$$

For a hypercubic lattice in the limit $k \to 0$, we find

$$\Delta^2(k) = 3j^2 a^2 k^2.$$

Similarly,

$$\begin{aligned}\langle \Omega_{kk}\Omega_{kk}\rangle_{AV} &= 2j^2\left(\sum_{j}(1 - e^{ik\cdot(r_i - r_j)})\right)^2\\ &= 2j^2 a^4 k^4,\end{aligned}$$

where the last line holds in the hydrodynamic limit.

3. Density of states For small k or energies, in three dimensions, we get

$$\rho(k) = \sum_{k'} \delta(\omega(k) - \omega(k')) = \frac{V}{(2\pi)^3}\frac{4\pi k^2}{d\omega/dk} = \frac{1}{(2\pi a)^2}\frac{k}{J_0}.$$

Using these results, we can obtain the k-dependence of the correlation time in the hydrodynamic limit, we get:

4. Correlation time

$$\tau_c(k) = \frac{V}{3\pi J_0}k^3.$$

5. Relaxation time

$$\Gamma(k) = \Delta^2(k)\tau_c(k) = \frac{aVj^2}{2\pi^2 J_0} k^5.$$

References

[1] Continentino M A and Rivier N 1979 *Phys. Status Solidi* B **93** 721

[2] Kubo R 1962 *J. Phys. Soc. Jpn.* **17** 1100
Kubo R 1962 *Fluctuation, Relaxation and Resonance in Magnetic Systems* ed D Ter Haar (London: Oliver and Boyd)

[3] Continentino M A and Rivier N 1977 *J. Phys. C: Solid State Phys.* **10** 3613

[4] Stanley H E 1987 *Introduction to Phase Transitions and Critical Phenomena* (New York: Oxford University Press)

[5] Edwards S F and Jones R C 1976 *J. Phys. A: Gen. Phys.* **10** 1595

IOP Publishing

Key Methods and Concepts in Condensed Matter Physics
Green's functions and real space renormalization group
Mucio A Continentino

Chapter 5

Real space renormalization group

5.1 Phase transitions and the renormalization group

Phase transitions are common phenomena in Nature; ice melts, water evaporates, some metals become superconductors at low temperatures and others develop a spontaneous magnetization, as we have seen in previous chapters. It was certainly a great success of physics to be able to understand, at least in part, these phenomena. The actual paradigm is due to Landau who introduced the fundamental concepts to treat the problem of phase transitions. The notion of an order parameter, symmetry breaking, critical exponents and the idea of expanding the free energy close to the phase transition in powers of the order parameter are most useful and laid the ground for amazing progress. We have obtained in chapters 3 and 4, the values for some critical exponents predicted by Landau theory. These exponents are known as mean-field since the Landau approach neglects fluctuations in the disordered phase and its results coincide with those obtained using mean-field approximations.

The ultimate test for a theory of critical phenomena is to predict values for the critical exponents that can be compared with experiments. As the quality of experiments and temperature control increased it became clear that the exponents predicted by Landau theory did not agree in many cases with experiments, or with those obtained in exact solutions of theoretical models.

Phase transitions can be divided in two large groups: those of first and those of second order depending whether the singular behavior is observed in quantities related to first or second derivatives of the free energy, respectively.

The breakthrough in the study of critical phenomena came with the work of Wilson, Kadanoff, and Fisher that culminated in the formulation of the renormalization group (RG) theory of second order phase transitions. A key concept in this approach is that of scale invariance. Physical quantities, close to a phase transition where fluctuations are large, depend on a single length scale that diverges at the critical point rendering the system scale invariant. The renormalization group associates the fixed point of a length scale transformation, a purely mathematical

doi:10.1088/978-0-7503-3395-5ch5

entity, with the critical point of the system. Under the change of length scales, physical quantities, like the parameters of the Hamiltonian are renormalized reaching definite values at the fixed points.

The most used and well-known version of the renormalization group is implemented in momentum space. This approach leads to values of the critical exponents in good agreement with experiments. However, the calculations of the RG equations in momentum space are elaborate and sometimes obscure the main ideas. On the other hand, the real space version of the RG that we present here [1] has the advantage of making the physical ideas transparent and consequently is very useful to introduce the basic concepts of the method. In addition the real space RG gives correct results for a class of systems known as *hierarchical lattices*. It has the deficiency to be uncontrolled, when used to describe systems in a real lattice. This means that increasing the size of the cells to obtain the RG equations does not necessarily improve the results for the critical exponents.

Below, we consider first the problem of percolation in a lattice. This is an interesting problem in itself and the real space RG provides a very useful approach. Without introducing the concept of an order parameter, we can fully characterize this problem just by considering the nature of the different fixed points and particularly of the attractors.

5.2 Bond percolation in a square lattice

Percolation refers to the movement of fluids in porous materials. As the number of pores increases, the fluid can eventually filter through the system, traversing it from one surface to the opposite. In condensed matter physics the problem of percolation is strongly connected with that of the appearance of long-range order in dilute magnetic systems, or the passage of electric current through a multi connected circuit.

Consider a square lattice where magnetic ions, which can interact with their first neighbors are placed at random. It is clear that if the system is very dilute there is no long-range magnetic order. The question, at which concentration of magnetic ions does long range magnetic order set in, is essentially a percolation problem. The appearance of long-range magnetic order requires that an *infinite* island of connected sites is formed in the material, which coincides with the percolation threshold.

A similar problem is that of a square lattice, for example, where conducting wires connect nearest neighbors lattice points. If these wires are removed at random, which fraction of them must remain, such that an electric current can flow through the material when the poles of a battery are connected to opposite sides?

Figure 5.1 represents a square lattice and highlighted in blue is a cell from which it can be formed [2]. The idea of the renormalization group is to transform this cell in a single bond that is going to be used to construct a new identical lattice, but with a different length scale. This new bond will be present in the new lattice depending on whether the cell from which it originates *percolates*, i.e., if current injected on the left extremity of the cell can flow and be collected at the right extremity. This procedure

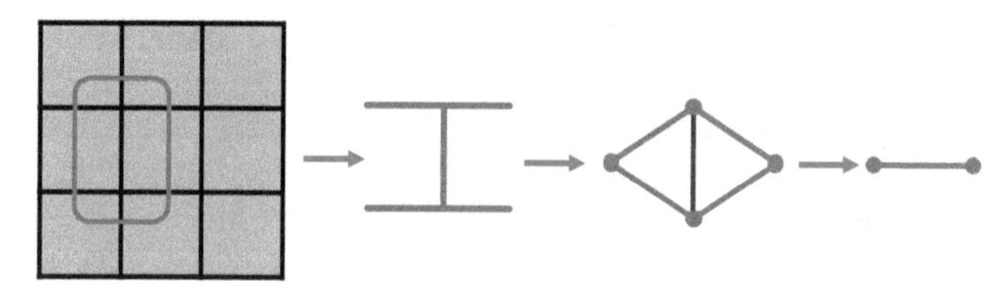

Figure 5.1. A square lattice of size L is shown in the left. In red is highlighted a small cell from which it can be reconstructed. For the study of percolation both left and both right terminals can be joined. Finally this cell is transformed into a single bond. This new bond will be present in the new lattice if the cell from which it originates percolates, i.e., it allows current injected in the left extremity to reach the right terminal. The renormalization of the last step corresponds to a scaling factor $b = 2$.

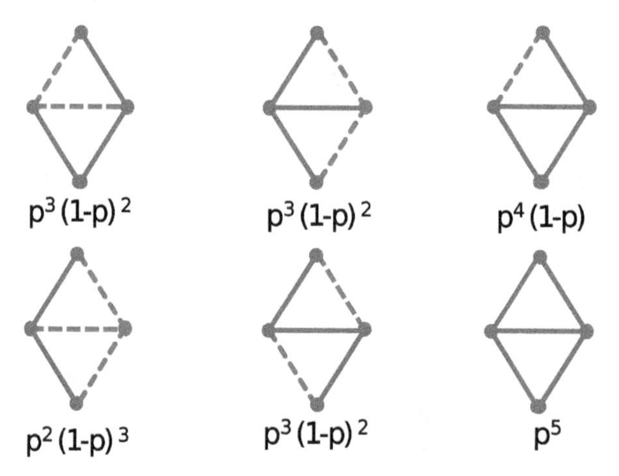

Figure 5.2. A percolating cell is that where a current can flow from the lower terminal to the upper one through existing bonds. The figures show some percolating cells and their respective probabilities. Dashed lines correspond to missing bonds.

is then repeated a number of times, i.e., the cells of the new lattice give rise to new bonds that form new effective lattices, until a fixed point is reached. Notice that what is renormalized in this case is the probability of a bond being present.

Figure 5.2 shows some percolating configurations and their respective probabilities. The probability of a bond being present is p and that of a missing bond $(1 - p)$. Figure 5.3 shows several non-percolating cells and their probabilities. For a total of five bonds in the cell there are $2^5 = 32$ possible configurations, but not all are percolating. The total number of percolating cells multiplied by their respective probabilities is given by,

$$f(p) = p^5 + 5p^4(1 - p) + 8p^3(1 - p)^2 + 2p^2(1 - p)^2. \tag{5.1}$$

The coefficients of the polynomial represent the number of cells with the same probabilistic weight. From the total of 32 cells only half of them are percolating. In the spirit of the RG the function $f(p)$ gives the probability of the renormalized bond

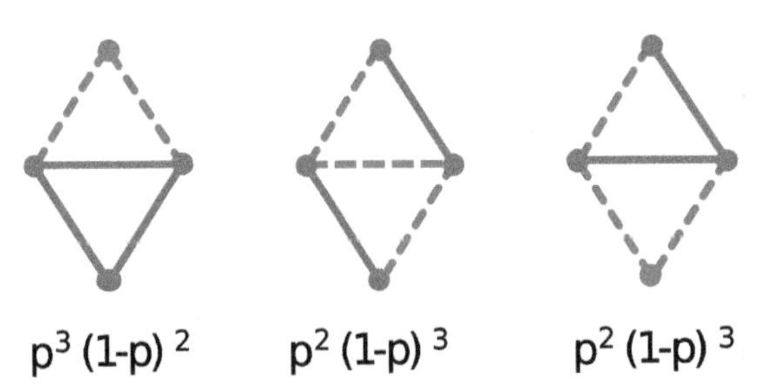

Figure 5.3. Examples of non-percolating cells and their respective probabilities. Dashed lines correspond to missing bonds.

being present in the scaled lattice. Consequently, the renormalization group equation for the probability of percolating configurations can be written as,

$$\begin{aligned} p' &= p^5 + 5p^4(1-p) + 8p^3(1-p)^2 + 2p^2(1-p)^2 \\ &= 2p^5 - 5p^4 + 2p^3 + 2p^2. \end{aligned} \tag{5.2}$$

The fixed points of this equation are those values of p for which $p^* = f(p^*)$. There three fixed points, $p_1 = 0$, $p_2 = 1/2$ and $p_3 = 1$. The easiest way to understand the nature of these fixed points is to write equation (5.2) as a recursion relation,

$$p_{n+1} = 2p_n^5 - 5p_n^4 + 2p_n^3 + 2p_n^2. \tag{5.3}$$

When we iterate these recursion relations, we obtain two different behaviors depending on the initial probability p_0. For $p_0 > p_2 = 1/2$ the RG equation iterates to the fixed point $p_3 = 1$, like in the sequence with $p_0 = 0.6$: (0.6, 0.659 52, 0.747 249, 0.858 283, 0.956 043, 0.995 984, 0.999 968, 1.). For $p_0 < p_2 = 1/2$, equation (5.3) iterates to $p_1 = 0$. For example taking $p_0 = 0.4$ we get: (0.4, 0.340 48, 0.252 751, 0.043 9574, 0.004 016 04, 0.000 032 3854, $2.097\,7 \times 10^{-9}$). Then the flow of the RG equation is always away from the *unstable* or *repulsive fixed point* p_2. It separates two totally different behaviors, the points that under renormalization flow to $p_1 = 0$ from those that flow to $p_3 = 1$. The fixed points $p_1 = 0$ and $p_3 = 1$ are *attractors* in the RG sense. These attractor fixed points, for which equation (5.3) iterates, inform us about the nature of the different phases of the system. When the system iterates to $p_1 = 0$, implying that all bonds are missing, using the analogy with the wired lattice, we can posit that the system is an insulator with no current flowing from one side to the other of the sample. On the other hand when the iteration is for $p_3 = 1$ all bonds are present and the system is clearly a conductor, and the current delivered in one side of the sample can be collected at the opposite side.

At the unstable fixed point $p_2 = 1/2$ separating these two behaviors there is a *percolation phase transition* that in the case of the wired lattice is associated with a metal–insulator transition. Thus, an unstable fixed point of a recursion relation has

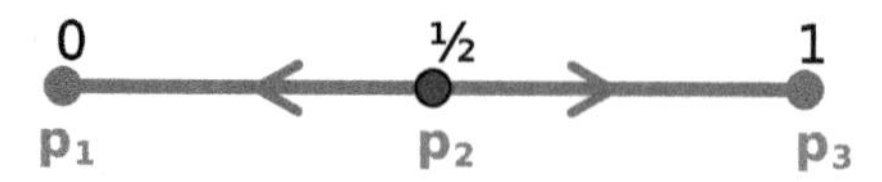

Figure 5.4. Flow diagram for percolation in a 2d lattice showing the stable fixed points p_1 and p_3 and the unstable fixed point at $p_2 = p_c = 1/2$.

been identified with the actual critical point of a phase transition in a physical system. This is one of the beauties of the RG. These results are illustrated in the phase diagram of figure 5.4. It shows all three fixed points and the flow of the RG equations. The fixed points p_1 (p_3) are attractors since for concentrations of bonds $p_0 < 1/2$ ($p_0 > 1/2$), respectively, the RG flow is towards them. Near the unstable fixed point $p_2 = 1/2$, all points p_0 tend to move away from it. This repulsive fixed point is that associated with the phase transition. The critical concentration of bonds $p_2 = p_c = 1/2$ is actually the exact value for bond percolation in a square lattice.

Near the unstable fixed point $p_2 = p_c = 1/2$, the RG equation can be expand as,

$$p_{n+1} = p_c + b^x(p_n - p_c), \tag{5.4}$$

where b is the length scale factor between the original and renormalized lattice ($b = L/L'$), in the present case $b = 2$. If one iterates equation (5.4) N times, we get

$$p_N = p_c + b^{Nx}(p_0 - p_c).$$

Taking $b^N = \ell$, where ℓ is a length and since the number N of iterations is arbitrary, we can repeat the RG procedure until,

$$\ell^x(p_0 - p_c) = 1$$

or

$$\ell = (p_0 - p_c)^{-1/x}.$$

We identify this length scale with the *correlation length* ξ of the system, which diverges at the phase transition and renders the system scale invariant. The exponent that controls this divergence is the *correlation length exponent* $\nu_P = 1/x$. For the purpose of determining this exponent, we rewrite equation (5.4) as,

$$p_1 - p_c = b^x(p_0 - p_c)$$

or

$$b^x = \frac{p_1 - p_c}{(p_0 - p_c)} = (dp'/dp)_{p=p_c},$$

where in the last equality we have taken the limit $p_0 \to p_c$. We finally obtain

$$\nu_P = \frac{\ln b}{\ln(dp'/dp)_{p=p_c}} \approx 1.43,$$

where we used that the length scale transformation used in the cells of figure 5.2 corresponds to $b = 2$. The exact value for the correlation length exponent in the square lattice is $\nu_P = 4/3 \approx 1.33$. The diverging length in the percolation problem can be identified with the size of the largest connected cluster in the lattice that becomes infinite at the percolation threshold.

5.3 Hierarchical lattices

Hierarchical lattices are fractal objects, with fractal dimensions. They do not have translation invariance as the pure systems without disorder that we have studied so far [1]. These lattices, for certain properties that we will explore are good approximations for periodic lattices of real systems. They are useful in the study of the critical properties and phase diagrams of real systems, especially in the presence of disorder.

In figure 5.5 we show some hierarchical lattices and their respective fractal dimensions D_f, defined as the logarithm of the total number of bonds in a cell divided by the logarithm of the minimum number of bonds to go from the bottom to the top of the cell. Notice that the form to generate these lattices is to replace each bond of the cell by the cell itself in an infinite process.

Since hierarchical lattices have no translation invariance, momentum k is not a good quantum number and they are better dealt in their own hierarchical space. From the point of view of critical phenomena these lattices have the advantage that the RG transformations, known as Migdal–Kadanoff transformations [1] are exact

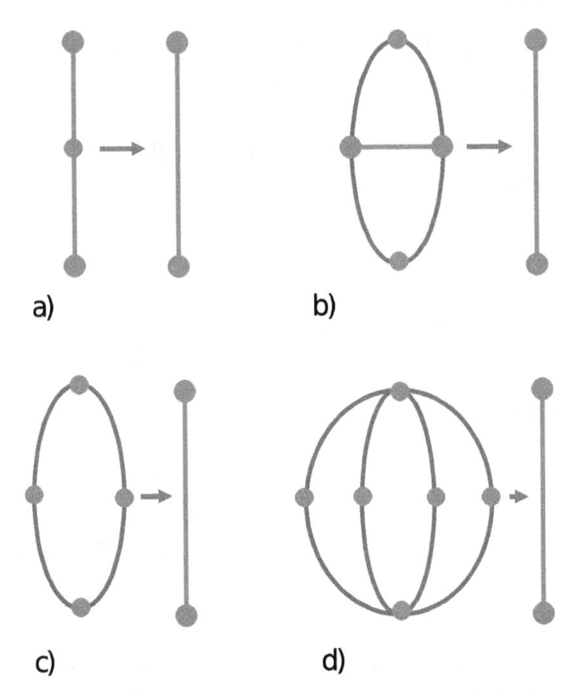

Figure 5.5. Hierarchical lattices and their fractal dimensions. (a) $D_f = \log 2/\log 2 = 1$, (b) $D_f = \log 5/\log 2$, (c) $D_f = \log 4/\log 2 = 2$ and (d) $D_f = \log 8/\log 2 = 3$. The arrows indicate the method of construction of the lattice where a single bond is replaced by the corresponding cell ad infinitum. See also figure 5.6.

in these systems. The RG recursion relations mimic the process of generation of these lattices.

For the study of critical phenomena, hierarchical lattices are useful approximations for real systems. However, they remain an approximation even if their fractal dimensions coincide with the real space dimension of the physical lattices, as in figure 5.5(a), (c) and (d). The advantage of implementing the RG in these lattices is its simplicity and that the complete parameter space can be mapped to show the flow of the RG equations, the fixed points and in particular the attractor. In some cases, the results obtained are very good when compared to exact solutions, or simulations of real physical systems. This holds especially for random systems, perhaps because both lack translational invariance.

Notice that when we solved the percolation problem in the square lattice, we used the diamond hierarchical lattice shown in figure 5.6 as an approximation. We obtained the correct result for the percolation threshold in spite of the fractal dimension of the diamond hierarchical lattice being $D_f = \ln 5/\ln 2 = 2.32$.

Notice, however, that the results for p_c and ν_P are exact for the diamond hierarchical lattice itself. This is plausible if we accept that the iteration of the RG equations mimics the process of construction of the hierarchical lattice itself.

In the next section, we will study the critical properties of the Ising model in hierarchical lattices.

5.4 The Ising model

The Ising model is a classical model of interacting spins that plays an important role in statistical mechanics. In particular, it can be exactly solved in a two-dimensional square lattice. This solution allows the determination of critical exponents that turn out to be different from those obtained in the theory of Landau. The Ising Hamiltonian is given by,

$$H = -\frac{1}{2}\sum_{ij} J_{ij} S_i^z S_j^z. \tag{5.5}$$

S_i^z represent classical spins that can take values $S^z = \pm 1$. The interactions J_{ij} couple pairs of nearest neighbors spins on sites i and j. This model is not rotational

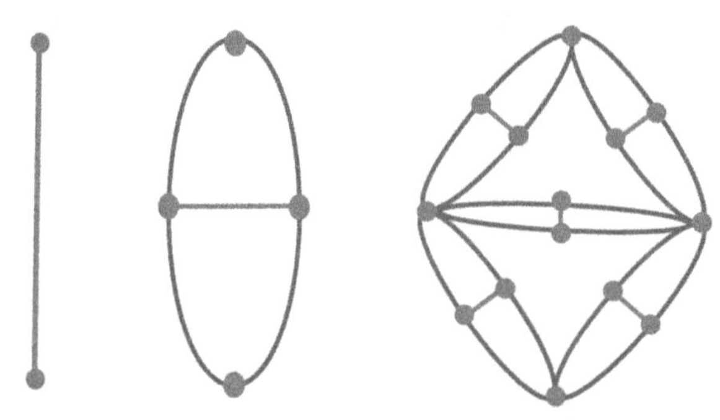

Figure 5.6. Three generations of the diamond hierarchical lattice, with fractal dimension $D_f = \ln 5/\ln 2 \approx 2.32$.

invariant, it does not have a Goldstone mode and has long-range magnetic order in two dimensions at finite temperature, differently from the Heisenberg model. However, it possesses a Z_2, up-down symmetry with a double degenerate ground state. Let us consider the Ising model on hierarchical lattices, such as those in figure 5.5. We are interested in obtaining the phase diagrams and the different phases at finite temperature. For this purpose we construct a quantity that is invariant under length scale transformations, i.e., it is the same in different generations of the lattice. We are going to consider the ratio of restricted partition functions where the spins at the terminals of a hierarchical cell are taken as fixed. This is defined by [3],

$$t = \frac{Z^{++} - Z^{+-}}{Z^{++} + Z^{+-}}, \tag{5.6}$$

where Z^{++} is the partition function of a cell, or bond, with the spins at both terminals pointing up. Z^{+-} is the partition function with the spin in the lower terminal pointing up and that in the upper terminal pointing down. This is clearly a scale invariant, since it is the *ratio* of partition functions. The quantity defined by equation (5.6) is known in the literature as *transmissivity* [3]. For a bond with a ferromagnetic interaction $J_{ij} = J > 0$ between the pair of spins, the transmissivity is given by,

$$t = \frac{Z^{++} - Z^{+-}}{Z^{++} + Z^{+-}} = \frac{e^{\beta J} - e^{-\beta J}}{e^{\beta J} + e^{-\beta J}} = \tanh \beta J, \tag{5.7}$$

where $\beta = 1/k_B T$ is the inverse of temperature.

These transmissivities have useful properties as they follow certain rules when we add different bonds, as can be directly verified. For bonds in series, as in figure 5.7(a), the total transmissivity is the product of the individual ones, $t_s = t_1 t_2$. For bonds in parallel as in figure 5.7(b), we have, $t_p = (t_1 + t_2)/(1 + t_1 t_2)$.

5.4.1 One-dimensional Ising model

The transformation in figure 5.7(a) represents the rescaling of the one-dimensional Ising model by a factor $b = 2$. According to the rule to add transmissivities in series, we have

$$t_{n+1} = t_{1n} t_{2n} = t_n^2$$

since the interactions are the same ($J_1 = J_2$).

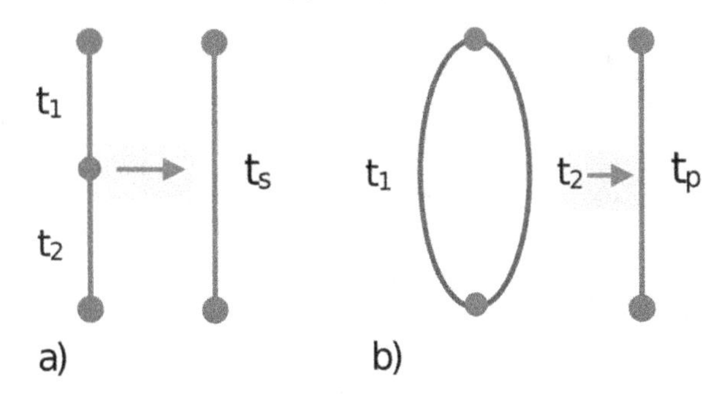

Figure 5.7. Sum of transmissivities in series and parallel.

The fixed points of this recursion relation are $t = 0$ and $t = 1$, which correspond to $(J/k_BT) = 0$ and $(J/k_BT) = \infty$, respectively, as shown in figure 5.8. Iterating the RG recursion relation we can verify that the former is an attractor and the latter a repeller. The attractor at $t = 0$ or $(J/k_BT) = 0$ is a *weak coupling fixed point* and is associated with paramagnetism and the absence of long-range order. Every point in the space of parameters J and T is attracted to this weak coupling fixed point. The consequence is that there is no long-range magnetic order in the one-dimensional Ising model for any finite temperature.

5.4.2 Two-dimensional Ising model

Consider the cell in figure 5.9, with fractal dimension $D_f = 2$. This lattice has bonds in series and in parallel. Let us first add the transmissivities in series to obtain two new effective transmissivities in parallel, we have,

$$t_1^* = t_1 t_3$$

and

$$t_2^* = t_2 t_4$$

Now using the rule for adding the transmissivities of bonds in parallel, we get

$$t^* = \frac{t_1^* + t_2^*}{1 + t_1^* t_2^*} = \frac{t_1 t_3 + t_2 t_4}{1 + t_1 t_2 t_3 t_4}.$$

Since the interactions in the bonds are equal, we obtain

$$t_{n+1} = \frac{2t_n^2}{1 + t_n^4}.$$

This equation has three real fixed points. Two of them are the same as the 1d case, namely, $t^* = 0$ and $t^* = 1$, *weak* and *strong coupling* fixed points, respectively. In this

t =0 t=1

Figure 5.8. Phase diagram of the one-dimensional Ising model showing the flow of the RG equation.

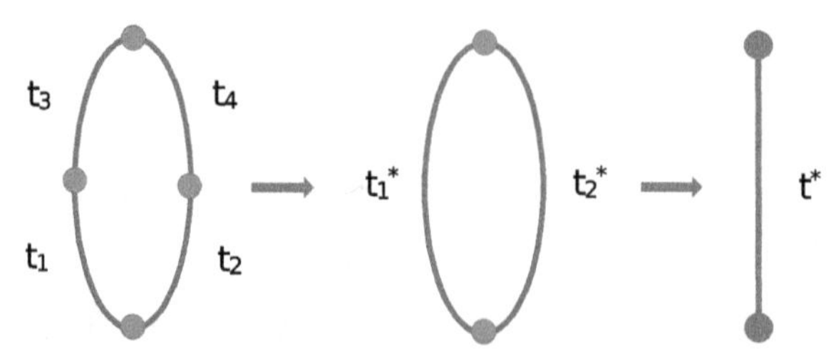

Figure 5.9. Cell with $D_f = 2$ showing renormalization of bonds in series and parallel. The transmissivities on the original cell are $t_i = \tanh(J_i/k_BT)$.

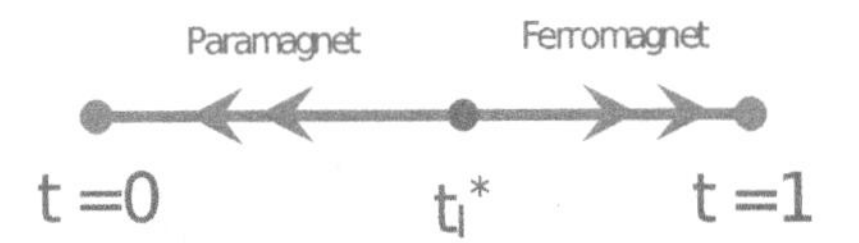

Figure 5.10. Phase diagram of the Ising model in a hierarchical lattice with $D_f = 2$. The unstable fixed point at t_I^* separates a paramagnetic from a long-range ordered ferromagnetic phase. The fixed points at $t = 0$ and $t = 1$ are the attractors of the paramagnet and ferromagnet, respectively.

case both of them are attractors. Since $t^* = 0$ implies $J/k_BT = 0$ this can be identified as the attractor of the paramagnetic phase where the spins are uncoupled. The strong coupling fixed point, that has exchanged its stability with respect to the 1d case, corresponds to $J/k_BT = \infty$. In this case the spins are strongly coupled and this attractor can be identified as that of a long range ordered ferromagnetic phase. Between these fixed points there is now a new fixed point at $t_I^* \approx 0.54$. This is an unstable fixed point. Points with $t > t_I^*$ will flow to strong coupling at $t^* = 1$ and those with $t < t_I^*$ flow to weak coupling at $t^* = 0$ under renormalization, as shown in figure 5.10. The unstable fixed point t_I^* is then clearly associated with a phase transition from a paramagnetic to a ferromagnetic phase. The critical temperature is obtained from $t_c = \tanh(J/k_BT_c) = 0.54$, that yields $k_BT_c \approx 1.66J$ compared to the exact result $k_BT_c \approx 2.27J$ for a square lattice. The critical exponent ν_T that controls the divergence of the correlation length at the magnetic phase transition, can be obtained as in the percolation problem and is given by,

$$\nu_T = \frac{\ln b}{\ln[(dt_{n+1}/dt_n)_{t_n=t_I^*}]}. \tag{5.8}$$

Since the scaling factor of the RG transformation, $b = 2$, we obtain $\nu_T = 1.34$. The exact solution of the Ising model on a square lattice yields, $\nu_T = 1$. Then the solution of the Ising model in a hierarchical lattice with fractal dimension $D_f = 2$ yields approximate results for the critical temperature and correlation length exponent of the 2d Ising model in a square lattice. However, these quantities are exact for the hierarchical lattice. A relevant concern is what happens if we try to improve the treatment above aiming to obtain better results for the two-dimensional Ising model in a square lattice. A possibility is to renormalize larger cells with the same fractal dimensions $D_f = 2$ and different scaling factors as shown in figure 5.11. A lot of work seems to confirm that this procedure does not necessarily lead to better results, leaving the notion that the real space RG is an uncontrolled approximation for real lattices.

5.4.3 Magnetic dilution

Let us consider the cell in figure 5.9 and start to dilute it by removing bonds between the magnetic ions [4].

We wish to know which is the critical concentrations of bonds at which magnetism disappears. We start by solving the percolation problem in this cell. There are a total of 2^4 configurations where bonds can be present with probability p and absent with probability $(1 - p)$. From all 16 possible configurations only seven

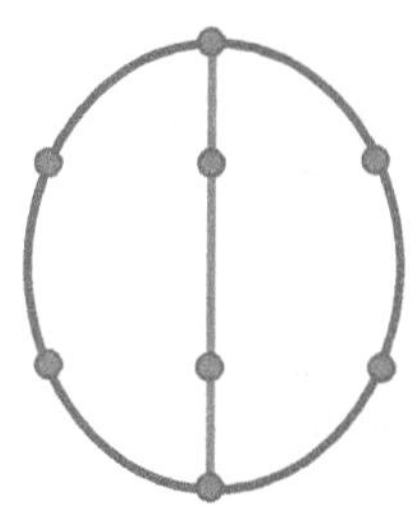

Figure 5.11. A hierarchical cell with fractal dimension $D_f = 2$ and a length scale $b = 3$.

are percolating. All configurations with three missing bonds are non-percolating. As before, we add the percolating cells multiplied by their probabilistic weight. When the cell is renormalized in a single bond, we obtain the following recursion relation,

$$p_{n+1} = p_n^4 + 4p_n^3(1 - p_n) + 2p_n^2(1 - p_n)^2. \tag{5.9}$$

The fixed points of this equation are the stable fixed points at $p = 0$ and $p = 1$ attractors of the *insulating and metallic phases*, respectively. There is also an unstable fixed point at $p_c \approx 0.62$ that separates these phases. The correlation length exponent can be easily obtained as in the previous case. We find $\nu_P = 1.63$.

Next, we imagine Ising spins occupying all sites in the lattice with the bonds representing ferromagnetic interactions between these spins. At which concentration of bonds does the ferromagnetic order disappear at zero temperature? This problem can be approached considering the transmissivities of all the percolating cells with their respective probabilistic weights. When renormalizing the $b = 2$ cell in a single bond as in figure 5.7, we obtain,

$$p_{n+1}t_{n+1} = p_n\frac{2t_n^2}{1 + t_n^4} + 4p_n^3(1 - p_n)t_n^2 + 2p_n^2(1 - p_n)^2t_n^2, \tag{5.10}$$

where p_{n+1} is the probability that this new bond, or interaction will be present in the next generation. Equations (5.9) and (5.10) have to be iterated together.

We now have two recursion relations and the information about the critical exponents is contained in the Jacobian matrix that arises linearizing the RG recursion relations close to the unstable fixed points,

$$\begin{pmatrix} \frac{\partial p'}{\partial p} & \frac{\partial p'}{\partial t} \\ \frac{\partial t'}{\partial p} & \frac{\partial t'}{\partial t} \end{pmatrix}_{p_c, t_c}.$$

Since $(\partial p'/\partial t) = 0$, the eigenvalues of this matrix are $(\partial p'/\partial p)$ and $(\partial t'/\partial t)$ evaluated at the fixed points. At the Ising fixed point $p = 1$, $t = t_c$, we get $(\partial t'/\partial t) = 1.679$ and $(\partial p'/\partial p) = 0$. From the first derivative we obtain the Ising correlation exponent $\nu_T = 1.34$, as in equation (5.8). The direction of the flow of a given *field* in parameter space close to a fixed point determines whether this field is *relevant* or *irrelevant*. In the former case the field iterates away from the fixed point and in the latter towards it. Mathematically, the condition for a field to be *relevant* along a given direction in phase space, specified by an eigenvector, is that the corresponding eigenvalue is larger than unity. In the case of bond dilution at the Ising fixed point, since $(\partial p'/\partial p) = 0$, we say dilution is a *marginal field*. In this case to determine the relevance, or irrelevance of dilution one has to calculate further terms in the expansion of the RG equations. We can easily verify by iterating the RG equations that starting from a finite value of the concentration of bonds p, close to the Ising fixed point this iterates towards $p = 1$, the pure system. Then dilution is an irrelevant perturbation close to the Ising fixed point. The physical implication is that the critical behavior, i.e., the critical exponents of the paramagnetic-to-ferromagnetic phase transition in a dilute system, are asymptotically the same as those of the pure Ising system, as long as, $p > p_c$.

There is a general criterion to decide whether fixed points are stable under disorder (dilution). The *Harris criterion* relies on the sign of the specific heat exponent α. For $\alpha < 0$ the stability of the fixed point is guaranteed and the contrary for $\alpha > 0$. For $\alpha = 0$ nothing can be said. In our case the exponent α can be obtained from the hyperscaling relation, which in the case of fractal lattices we write as $2 - \alpha = \nu D_f$. Using the value of $\nu_T = 1.337$ obtained above for the Ising fixed point and $D_f = 2$, we get $\alpha = -0.674$, consistent with the stability of this fixed point under dilution as obtained above.

Figure 5.12 shows the t versus p phase diagram of the hierarchical lattice studied above. There are three phases in the system,

- The percolating paramagnet above the dashed line at $p = p_c$ in figure 5.12. All the points in the phase diagram belonging to this phase iterate under the RG equations to the attractor fixed point at $p = 1$, $t = 0$.
- In the region above the line $p = p_c$, there is a critical line separating the percolating paramagnet from the long-range ordered ferromagnetic phase. The points inside the ferromagnetic phase iterate to the strong coupling attractor of the pure Ising ferromagnet at $p = 1$, $t = 1$.
- Finally, below the dashed line $p = p_c$, all the points in this region, excluding those at the line $t = 1$, iterate to the trivial attractor at $p = 0$, $t = 0$.

Figure 5.13 shows the same phase diagram but now on the physical variables, temperature (T) versus concentration of bonds p. There is a line $T_c(p)$ that, for $p > p_c$ separates the ferro and paramagnetic phases. The arrows indicate the flow of the RG equations close to the unstable fixed points and along the line $T_c(p)$. The irrelevance of disorder (dilution) close to the Ising unstable fixed point is indicated by the arrow showing the direction of the flow along the critical line towards this fixed point. The paramagnetic-to ferromagnetic phase transitions along this line are all asymptotically governed by the critical exponents of the pure system.

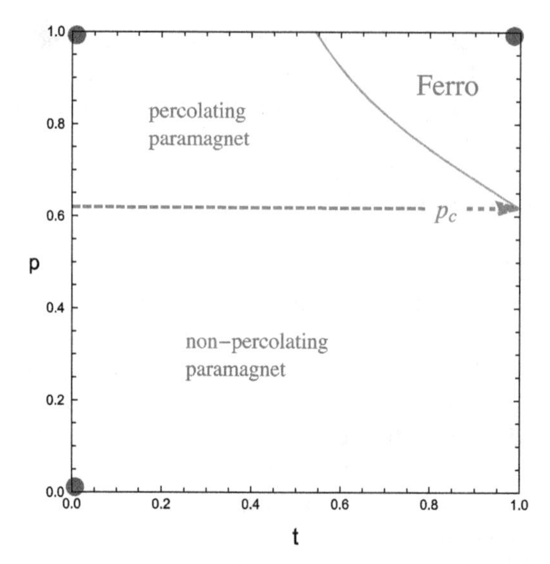

Figure 5.12. Phase diagram p versus t of the Ising model in a hierarchical cell with $D_f = 2$. The line separating the ferromagnetic phase from the percolating paramagnet is the *critical line*. It separates the points in the plane that flow under the recursion relations to the strong coupling attractor of the pure system at $t = 1, p = 1$ from those that iterate to $t = 0, p = 1$. The points below the dashed red line away, excluding the line $t = 1$, iterate to the trivial attractor at $t = 0, p = 0$.

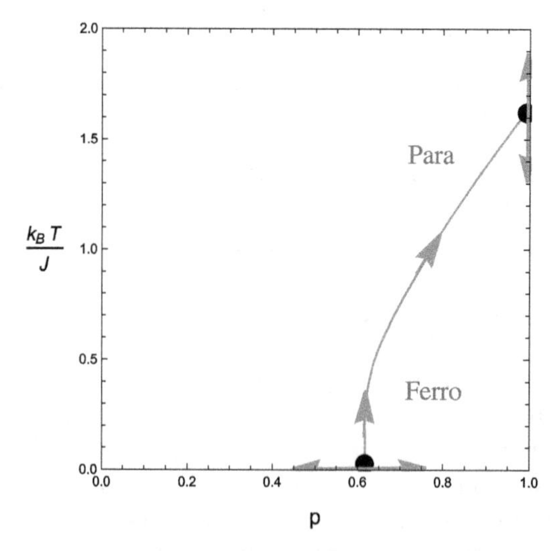

Figure 5.13. The same phase diagram of figure 5.12 plotted in the physical variables $T \times p$. The critical line $T_c(P)$ separates the ferromagnet from the percolating paramagnet. The arrows indicate the flow of the RG recursion relations, equations (5.9) and (5.10). Any point $p_0 > p_c$ on the critical line iterates to the pure Ising fixed point showing the irrelevance of disorder (dilution).

Finally, we emphasize the close connection between percolation and magnetism. It is clear from the phase diagrams in figure 5.12 and 5.13 that long-range magnetic order can only exist once an infinite cluster of bonds percolates the material. In other words $T_c(p)$ is finite only for $p > p_c$.

5.5 Ising ferromagnet in a magnetic field

The Hamiltonian that describes the Ising ferromagnet in a uniform magnetic field h in a hierarchical lattice is given by,

$$H = -\frac{1}{2}\sum_{ij} J_{ij}S_i^z S_j^z - (h/2)\sum_i z_i S_i^z, \tag{5.11}$$

where the quantity z_i is a weight in the one-body Zeemann term equal to the number of bonds leaving the site i. It is necessary to compensate for the difference of coordination number of the sites in the cells [5]. Let us consider the effect of the magnetic field in the pure (without dilution) hierarchical lattice with $D_f = 2$ studied in the previous section. We use the transmissivities to obtain the phase diagram. Since the Z_2 symmetry has been broken by the external magnetic field, we have now that $Z^{++} \neq Z^{--}$. This allows us to define another invariant that will provide a recursion relation for the magnetic field. The relevant transmissivities are,

$$\begin{aligned} t^+ &= \frac{Z^{++} - Z^{+-}}{Z^{++} + Z^{+-}} \\ t^- &= \frac{Z^{--} - Z^{-+}}{Z^{--} + Z^{-+}}. \end{aligned} \tag{5.12}$$

For a single bond these quantities are easily calculated and we get,

$$\begin{aligned} t^+ &= \tanh \beta(J + h/2) \\ t^- &= \tanh \beta(J - h/2). \end{aligned} \tag{5.13}$$

Next step is to calculate these quantities for the same cell of the previous section with $D_f = 2$.

In figure 5.14 we show all spin configurations used to calculate Z^{++}. The spins on the upper and lower terminal are fixed and pointing up in the direction of the external field. We find

$$Z^{++} = e^{4\beta(J+h)} + e^{2\beta h} + e^{2\beta h} + e^{-4\beta J},$$

where each term corresponds to a given spin configuration. Notice that in this case all the sites have the same weight $z_i = 2$ since all are connected to two bonds. In the same way we obtain the other restricted partition functions, Z^{+-}, Z^{-+} and Z^{--}. We recall that the notation, Z^{+-} refers to the spins on the upper and lower terminals fixed and pointing up (+) and down (−), respectively. After a long but straightforward calculation we get for the transmissivities of the cell,

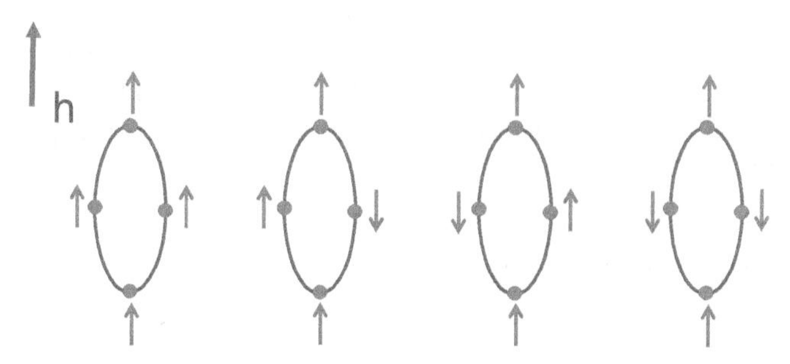

Figure 5.14. Configurations of the cells used to calculate Z^{++}. The external field h points in the $\mathbf{z}$ direction.

$$\begin{aligned} t^+ &= \frac{e^{4\beta(J+h)} + e^{-4\beta J} + e^{2\beta h} - e^{-2\beta h} - 2}{e^{4\beta(J+h)} + e^{-4\beta J} + 3e^{2\beta h} + e^{-2\beta h} + 2} \\ t^- &= \frac{e^{-4\beta J} + e^{4\beta(J-h)} + e^{-2\beta h} - e^{2\beta h} - 2}{e^{-4\beta J} + e^{4\beta(J-h)} + 3e^{-2\beta h} + e^{2\beta h} + 2}. \end{aligned} \tag{5.14}$$

The RG recursion relations are obtained equating equations (5.13) and 5.14. For zero external magnetic field ($h = 0$), both recursion relations coincide with that of the pure ferromagnet ($p = 1$) of the previous section and the phase diagram is the same. Since the magnetic field is conjugate to the order parameter there is no phase transition in the presence of this field. Then, for any finite field the flow of the recursion relations is to the paramagnetic fixed points $h = \pm\infty$, $J = 0$, depending on the sign of the field, as shown in figure 5.15. In terms of the transmissivities the flow is to $(t^+, t^-) = (1, -1)$ for $h > 0$ and $(-1, 1)$ for $h < 0$.

In figure 5.15, I is the fully unstable Ising fixed point at $h = 0$, $T = T_c$, with $T_c/J = 1.641\,02$. The fixed point StC is the strong coupling fixed point, attractor of the ferromagnetic phase. The line $h = 0$ is a line of first order transitions. The flows of the RG equations are also shown in this figure. Since there are two relevant *fields* at the Ising fixed point, namely temperature and magnetic field, the relevant information on the critical exponents is obtained from the matrix,

$$\begin{pmatrix} \frac{\partial h'}{\partial h} & \frac{\partial h'}{\partial T} \\ \frac{\partial T'}{\partial h} & \frac{\partial T'}{\partial T} \end{pmatrix}_I = \begin{pmatrix} 3.678\,57 & 0 \\ -2.989\,76 \times 10^{-16} & 1.678\,57 \end{pmatrix}.$$

This has two eigenvalues with orthogonal eigenvectors, as shown in figure 5.15. The thermal eigenvalue along the temperature axis is $\lambda_T = 1.678\,57$. The field eigenvalue in the orthogonal direction is $\lambda_h = 3.678\,57$. The correlation length exponent,

$$\nu = \ln 2/\ln \lambda_T = 1.34.$$

We can also obtain the *crossover exponent* $\phi = \lambda_h/\lambda_T = 2.52$.

The scaling form of the free energy close to the Ising fixed point can be written as,

$$f = |t|^{2-\alpha} F\left[\frac{h}{|t|^{\phi}}\right], \tag{5.15}$$

where $t = T - T_c$ is the distance to the critical point and $F[x]$ a scaling function.

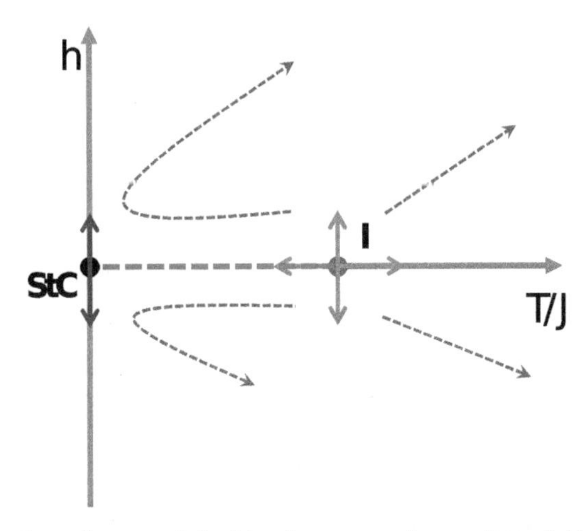

Figure 5.15. Schematic phase diagram of the Ising ferromagnet in a uniform field ($D_f = 2$). **I** is the fully unstable Ising fixed point at $T = T_c$, $h = 0$. The dashed red line is a line of first order transitions. **StC** is the strong coupling fixed point, attractor of the ferromagnetic phase. At this fixed point the external field scales as $h' = b^d h$ revealing the first order character of the transition line. The arrows show the flow of the RG equations.

The magnetization $m \propto (\partial f/\partial h)_{h=0}$ vanishes at the critical point as $m \propto |t|^\beta$ with the critical exponent β. The divergence of the susceptibility $\chi = (\partial^2 f/\partial h^2)_{h=0} \propto |t|^{-\gamma}$ defines the critical exponent γ. Calculating the derivatives of the free energy and assuming that $F'[0]$ and $F''[0]$ are finite, we obtain, the relations, $2 - \alpha - \phi = \beta$ and $2 - \alpha - 2\phi = -\gamma$, which in turn yield $\phi = \beta + \gamma$. Finally, the magnetization at the critical point in an applied magnetic field defines the critical exponent δ through the relation $m \propto h^{1/\delta}$ at $T = T_c$, $h \to 0$. A well-defined behavior for the magnetization in this limit requires that the scaling function $F'(x)$ behaves as a power law, such that

$$m \propto (\partial f/\partial h) \propto |t|^\beta \left(\frac{h}{|t|^\phi}\right)^x$$

as $h \to 0$. The requirement of a finite result requires that the dependence on t cancels out and consequently, $\phi x = \beta$ or $x = \beta/\phi$. Since $m \propto h^{1/\delta} = h^{\beta/\phi}$ in this limit, we obtain the exponents relation, $\delta = \phi/\beta$. Then assuming the validity of the hyper-scaling relation, $2 - \alpha = \nu D_f$, where $D_f = 2$ is the fractal dimension of the hierarchical lattice, and using the results above for ν and ϕ, we obtain the critical exponents of this lattice. In table 5.1 we list these critical exponents and compare them with the exact results for the Ising model in a square lattice.

5.6 First order phase transitions

As we pointed out, the red dashed line in figure 5.15 represents a line of first order transitions. The flow along this line is towards the strong coupling attractor $T/J = 0$, $h = 0$. Is there any specific feature of this attractive fixed point that allows us to

Table 5.1. The Ising exponents of the ferromagnetic transition for a hierarchical and a square lattice in two dimensions.

Ising model critical exponents		
	Hierarchical lattice $D_f = 2$	Square lattice
α	−0.68	0
β	0.16	1/8
γ	2.36	7/4
δ	15.75	15
ν	1.34	1
ϕ	2.52	1.875

identify this line, unambiguously, as one of discontinuous transitions? This important question was clarified by the work of Nienhuis and Nauenberg [6] showing that a definite signature of the first order character of this line is the existence of an eigenvalue at the strong coupling fixed point that corresponds to the dimensionality of the system. Since the magnetic field has to reverse the whole volume of a block of spins as it changes sign, the energy of this process scales as the volume of the system. This requires that the field scales as $h' = b^d h$ at the strong coupling fixed point where the spins are strongly coupled together. The real space RG recursion relations given above can be expanded near the StC attractor at $T/J = 0$, $h = 0$ and we obtain the recursion relations, $h' = 4h$ and $J' = 2J$. These are in agreement with the expected scaling of the conjugate field and the magnetic interaction of an Ising system close to this strong coupling attractor, namely,

$$\begin{cases} J' = b^{d-1}J & \text{for} \quad h = 0 \\ h' = b^d h, & \end{cases} \tag{5.16}$$

where the length scale factor $b = 2$ and we identify d with the fractal dimension of the cell, $d = D_F = 2$. The first equation in equation (5.16) for the coupling is characteristic of an Ising system for which the interaction between spin aligned clusters is through the surface of these clusters. For completeness we remark that for a Heisenberg system close to StC is $J' = b^{d-2}$ since in this case the domain walls are not sharp.

References

[1] Burkhardt T W and van Leeuwen J M J (ed) 1882 *Real-space RenormalizationTopics in Current Physics* vol 30 (Berlin, Heidelberg: Springer)

[2] de Oliveira P M C 1983 *Ciência Hoje* **2** 17

[3] Tsallis C and Levy S V F 1980 *J. Phys. C: Solid State Phys.* **13** 465

[4] Stinchcombe R B 1979 *J. Phys. C: Solid State Phys.* **12** 4533
Yeomans J M and Stinchcombe R B 1979 *J. Phys. C: Solid State Phys.* **12** 347
Stinchcombe R B 1983 *Phase Transitions and Critical Phenomena* vol 7 ed C Domb and J L Lebowitz (New York: Academic)
[5] Oliveira S M, Oliveira P M and Continentino M A 1988 *Physica* A **152** 477
[6] Nienhuis B and Nauenberg M 1975 *Phys. Rev. Lett.* **35** 477

IOP Publishing

Key Methods and Concepts in Condensed Matter Physics
Green's functions and real space renormalization group
Mucio A Continentino

Chapter 6

Real space renormalization group: quantum systems

6.1 Quantum systems

The real space renormalization group (RG) has also been used to treat quantum systems. However, in this case it is more difficult to implement. The main problem in this case is due to the non-commutativity of the terms in the Hamiltonian. In this chapter we discuss different approaches using the real space RG in quantum systems. In many cases they provide solutions consistent with the known lower critical dimension of the models. We start with the treatment of Takano and Suzuki of the spin-1/2 Heisenberg model in hierarchical cells of arbitrary dimensions. We will consider hierarchical cells of the type shown in figures 5.5(c), (d) and 6.1. In the quantum case, it can be easily verified that bonds in **parallel** just add [1]. For two bonds J_1 and J_2 in parallel the effective bond is just the sum of the bonds, i.e., $J_{12} = J_1 + J_2$. The main difficulty resides in the calculation of the effective bonds when these are in series. In this case the Heisenberg Hamiltonian describing three spin-1/2 in series is given by,

$$-\beta\mathcal{H}_{123} = J\mathbf{S}_1 \cdot \mathbf{S}_3 + J\mathbf{S}_3 \cdot \mathbf{S}_2, \tag{6.1}$$

where $\beta = 1/k_BT$, $\mathbf{S} = \sigma/2$, with σ (σ_x, σ_y, σ_z) the Pauli matrices and we considered identical interactions J. The renormalized interaction J' between spins 1 and 2 is obtained from the equation,

$$e^{(G'+J'\mathbf{S}_1\cdot\mathbf{S}_2)} = Tr_{\mathbf{S}_3} e^{J(\mathbf{S}_1\cdot\mathbf{S}_3+J\mathbf{S}_3\cdot\mathbf{S}_2)}, \tag{6.2}$$

where the trace operation is over the states of the intermediary spin $\mathbf{S}_3$. The constant G' is required to obtain meaningful equations. The implementation of the method has the following steps [2]:

doi:10.1088/978-0-7503-3395-5ch6

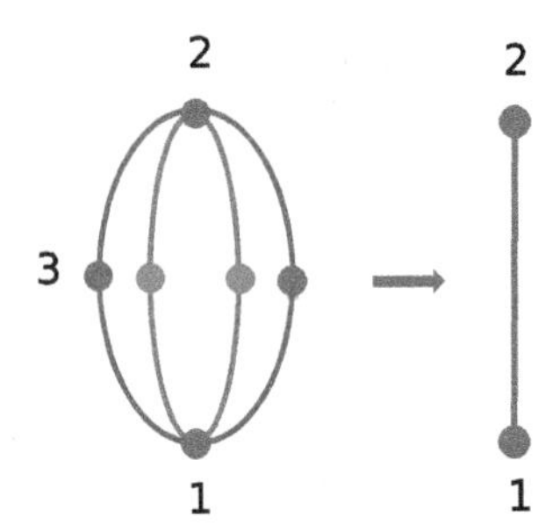

Figure 6.1. The *onion* cell with fractal dimension $D_f = \ln 8/\ln 2 = 3$ with a length scale transformation with factor $b = 2$.

- The left hand side of equation (6.2) is expanded as

$$e^{\mathcal{H}'_{12}} = e^{(G'+J'\mathbf{S}_1\cdot\mathbf{S}_2)} = a' + b'_{12}(\mathbf{S}_1 \cdot \mathbf{S}_2) \tag{6.3}$$

where the coefficients a' and b'_{12} depend on G' and J'.
- The right hand side of equation (6.2) yields,

$$\begin{aligned} e^{\mathcal{H}_{132}} = e^{J(\mathbf{S}_1\cdot\mathbf{S}_3+\mathbf{S}_3\cdot\mathbf{S}_2)} &= a + b_{13}(\mathbf{S}_1 \cdot \mathbf{S}_3) + b_{32}(\mathbf{S}_3 \cdot \mathbf{S}_2) \\ &+ b_{12}(\mathbf{S}_1 \cdot \mathbf{S}_2), \end{aligned} \tag{6.4}$$

where the coefficients are functions of J.
- Finally, taking the trace over $\mathbf{S}_3$ on the previous equation and using equations (6.2), (6.3) and (6.4), we obtain the relations,

$$\begin{cases} a' &= 2a \\ b'_{12} &= 2b_{12} \end{cases} \tag{6.5}$$

that relates J' in the renormalized cell (bond) with J in the three sites cell. Since $\mathbf{S}_3$ is traceless, the operation of taking the trace avoids the proliferation of new terms on the right hand side of equation (6.4).

The next step is to determine the a' and b' in terms of J' and G' and the a and b as functions of J and G. The way to proceed is to obtain unitary transformations U that diagonalize the matrices $\mathcal{H}'_{12}$ and $\mathcal{H}_{132}$, i.e., $U^{+}\mathcal{H}U = \mathcal{H}_D$, where $\mathcal{H}_D$ is the diagonal form of $\mathcal{H}$, or alternatively, $\mathcal{H} = U\mathcal{H}_D U^{+}$. Exponentiating both sides of this equation one gets, $\exp(\mathcal{H}) = U\exp(\mathcal{H}_D)U^{+}$ since $UU^{+} = 1$.

Let us show how this works for the two spins cell. The two spins, 4×4 matrix Hamiltonian can be easily diagonalized and the unitary transformation U determined. We get for $\exp(\mathcal{H}'_{12}) = U\exp(\mathcal{H}'_{12D})U^{+}$,

$$\exp\mathcal{H}'_{12} = \begin{pmatrix} e^{\epsilon_1} & 0 & 0 & 0 \\ 0 & \frac{1}{2}(e^{\epsilon_2} + e^{\epsilon_1}) & \frac{1}{2}(-e^{\epsilon_2} + e^{\epsilon_1}) & 0 \\ 0 & \frac{1}{2}(-e^{\epsilon_2} + e^{\epsilon_1}) & \frac{1}{2}(e^{\epsilon_2} + e^{\epsilon_1}) & 0 \\ 0 & 0 & 0 & e^{\epsilon_1} \end{pmatrix}$$

where $\epsilon_1 = G + J'$ and the triple degenerate eigenvalue $\epsilon_2 = G - 3J'$.

Equation (6.3) can be expressed in matrix form as,

$$\exp \mathcal{H}'_{12} = \begin{pmatrix} a' + b'_{12} & 0 & 0 & 0 \\ 0 & -a' + b'_{12} & 2a' & 0 \\ 0 & 2a' & -a' + b'_{12} & 0 \\ 0 & 0 & 0 & a' + b'_{12} \end{pmatrix}.$$

Comparing terms in the matrices, we get,

$$\begin{cases} e^{4J'} & = \dfrac{a' + b'_{12}}{a' - 3b'_{12}} = \dfrac{a + b_{12}}{a - 3b_{12}} \\ e^{G'} & = \dfrac{a' + b'_{12}}{e^{J'}} = \dfrac{2(a + b_{12})}{e^{J'}} \end{cases} \tag{6.6}$$

where the last equalities are consequences of equations (6.5).

In order to treat the three-spin cells, it is useful to introduce the following notation,

$$\begin{cases} S_{1\alpha} & = \dfrac{1}{2}\sigma_\alpha \otimes \sigma_0 \otimes \sigma_0 \\ S_{3\alpha} & = \dfrac{1}{2}\sigma_0 \otimes \sigma_\alpha \otimes \sigma_0 \\ S_{2\alpha} & = \dfrac{1}{2}\sigma_0 \otimes \sigma_0 \otimes \sigma_\alpha \end{cases}$$

where σ_α are the three Pauli matrices ($\alpha = x, y, z$), σ_0 is the 2×2 identity matrix and $\otimes$ stands for the Kronecker product. In this notation the three spins Hamiltonian, equation (6.1) can be written as

$$\begin{aligned} \mathcal{H}_{132} = & J(S_{1x} \cdot S_{3x} + S_{1y} \cdot S_{3y} + S_{1z} \cdot S_{3z}) \\ & + J(S_{3x} \cdot S_{2x} + S_{3y} \cdot S_{2y} + S_{3z} \cdot S_{2z}), \end{aligned} \tag{6.7}$$

where the $(\cdot)$ represents the usual matrix product. This 8×8 Hamiltonian can be diagonalized and the unitary matrix U obtained. The three-spins equivalent of equation (6.1) can be written as,

$$\exp \mathcal{H}_{132} = \tag{6.8}$$

$$\begin{pmatrix} e^{\epsilon_2} & 0 & 0 & 0 & 0 & 0 & 0 & 0 \\ 0 & \frac{1}{6}(e^{\epsilon_1} + 2e^{\epsilon_2} + 3) & \frac{1}{3}e^{\epsilon_1}(e^{3\epsilon_2} - 1) & 0 & \frac{1}{6}(e^{\epsilon_1} + 2e^{\epsilon_2} - 3) & 0 & 0 & 0 \\ 0 & \frac{1}{3}e^{\epsilon_1}(e^{3\epsilon_2} - 1) & \frac{1}{3}e^{\epsilon_1}(e^{3\epsilon_2} + 2) & 0 & \frac{1}{3}e^{\epsilon_1}(e^{3\epsilon_2} - 1) & 0 & 0 & 0 \\ 0 & 0 & 0 & \frac{1}{6}(e^{\epsilon_1} + 2e^{\epsilon_2} + 3) & 0 & \frac{1}{3}e^{\epsilon_1}(e^{3\epsilon_2} - 1) & \frac{1}{6}(e^{\epsilon_1} + 2e^{\epsilon_2} - 3) & 0 \\ 0 & \frac{1}{6}(e^{\epsilon_1} + 2e^{\epsilon_2} - 3) & \frac{1}{3}e^{\epsilon_1}(e^{3\epsilon_2} - 1) & 0 & \frac{1}{6}(e^{\epsilon_1} + 2e^{\epsilon_2} + 3) & 0 & 0 & 0 \\ 0 & 0 & 0 & \frac{1}{3}e^{\epsilon_1}(e^{3\epsilon_2} - 1) & 0 & \frac{1}{3}e^{\epsilon_1}(e^{3\epsilon_2} + 2) & \frac{1}{3}e^{\epsilon_1}(e^{3\epsilon_2} - 1) & 0 \\ 0 & 0 & 0 & \frac{1}{6}(e^{\epsilon_1} + 2e^{\epsilon_2} - 3) & 0 & \frac{1}{3}e^{\epsilon_1}(e^{3\epsilon_2} - 1) & \frac{1}{6}(e^{\epsilon_1} + 2e^{\epsilon_2} + 3) & 0 \\ 0 & 0 & 0 & 0 & 0 & 0 & 0 & e^{\epsilon_2} \end{pmatrix}$$

where the eigenvalues $\epsilon_1 = -4J$ and $\epsilon_2 = 2J$. The former is double degenerate and the latter has a degeneracy of 4. There are also two zero eigenvalues.

The expansion of equation (6.4), yields

$$\exp \mathcal{H}_{132} = \tag{6.9}$$

$$\begin{pmatrix} a+b_{12}+2b_{13} & 0 & 0 & 0 & 0 & 0 & 0 & 0 \\ 0 & a-b_{12} & 2b_{13} & 0 & 2b_{12} & 0 & 0 & 0 \\ 0 & 2b_{13} & a+b_{12}-2b_{13} & 0 & 2b_{13} & 0 & 0 & 0 \\ 0 & 0 & 0 & a-b_{12} & 0 & 2b_{13} & 2b_{12} & 0 \\ 0 & 2b_{12} & 2b_{13} & 0 & a-b_{12} & 0 & 0 & 0 \\ 0 & 0 & 0 & 2b_{13} & 0 & a+b_{12}-2b_{13} & 2b_{13} & 0 \\ 0 & 0 & 0 & 2b_{12} & 0 & 2b_{13} & a-b_{12} & 0 \\ 0 & 0 & 0 & 0 & 0 & 0 & 0 & a+b_{12}+2b_{13} \end{pmatrix}$$

where we took $b_{13} = b_{32}$.

Comparing terms of the matrices, equations (6.8) and (6.9), we obtain,

$$\begin{cases} 2b_{12} = \dfrac{1}{6}(e^{\epsilon_1} + 2e^{\epsilon_2} - 3) \\ a - b_{12} = \dfrac{1}{6}(e^{\epsilon_1} + 2e^{\epsilon_2} - 3) \end{cases}.$$

Solving for a and b_{12} and substituting in equations (6.6), we obtain the recursion relations,

$$\begin{cases} e^{4J'} = \dfrac{1}{3}(e^{-4J} + 2e^{2J}) \\ e^{G'} = 2e^{3J'} \end{cases} \tag{6.10}$$

or

$$J' = \frac{1}{4}\ln\left(\frac{1}{3}(2e^{2J} + e^{-4J})\right) \tag{6.11}$$
$$G' = \ln 2 + 3J'.$$

These are the recursion relations for the one-dimensional Heisenberg model. The fixed points of the interaction are at $J^* = 0$ ($T = \infty$) and $J = \infty$ ($T = 0$). The latter is unstable and the former is the weak coupling or high temperature attractor of the paramagnetic phase. This result is fully consistent with the well-known absence of long-range magnetic order in the 1d Heisenberg model.

Now that we have calculated the effective interaction for two interactions in series and since interactions in parallel just add up, we can obtain general recursion relations for cells of arbitrary dimensions with a length scale factor $b = 2$,

$$J' = \frac{b^{d-1}}{4}\ln\left(\frac{1}{3}(2e^{2J} + e^{-4J})\right). \tag{6.12}$$

We can easily check that for $d = 2$, the fixed points are the same as in the one-dimensional case. The absence of a non-trivial fixed at a finite value of J indicates that there is no finite temperature transition in this model in $d = 2$, in agreement with the spin-wave result of chapter 2. As pointed out before, this is a well-known result implied by the Mermin–Wagner theorem. In three dimensions there is now a non-trivial fixed point at $J^* = 0.34$. This fixed point is fully unstable and is associated with the ferromagnetic transition of the model. As in the Ising case, we can obtain the correlation length exponent expanding the recursion relations close to this fixed point.

If we consider an expansion of the RG equations close to the strong coupling fixed point, i.e., the attractor of the ferromagnetic phase at $J = \infty$, for $D_f = 3$, we find that the interaction scales as $J' = 2J$. This is the expected behavior for a model of continuous symmetry close to this fixed point, i.e., $J' = b^{d-2}J$. This exponent is due to the finite width of the domain walls separating islands of aligned spins, different from the abrupt changes in Ising systems with discrete symmetry.

According to Mermin–Wagner, two-dimensions is the *lower critical dimensional* for the Heisenberg model. This means that only for $d > 2$ is there a finite temperature transition in this model with a continuous symmetry. This can be used to perform an expansion close to two dimensions. Let us consider equation (6.12) with $d = 2 + \epsilon$. Expanding close to $T = 0$ and for small ϵ, we find an unstable fixed point at $T^* = 1/J^* = -\epsilon(2\ln 2/\ln(2/3)) = (3.42)\epsilon$ that vanishes as $\epsilon \to 0$. Notice that this type of expansion is different from the usual ϵ-expansion where one expands the RG equations close to the *upper-critical dimension* above which the exponents become mean-field.

6.2 The free energy

So far we have used the real space RG to obtain phase diagrams, flows in parameter space, fixed points and critical exponents. However, this approach is more powerful and allows us to obtain thermodynamic quantities like, specific heat, magnetization, even far away from the critical points [1].

These calculations rely on recursion relations for the free energy from which the relevant physical quantities can be derived. The recursion relation for the free energy corresponding to the cell renormalization we used in the previous section is given by,

$$G + F' = \ln \sum_{\sigma'} e^{G+\mathcal{H}'(\sigma')} = \ln \sum_{\sigma} \mathrm{Tr}_{\sigma'} e^{\mathcal{H}(\sigma', \sigma)}$$
$$= \ln \sum_{\sigma} e^{\mathcal{H}(\sigma)} = F.$$

Since the free energy is an extensive quantity, we have $F = Nf(J)$, $G = Ng(J)$ and $F' = N'f(J')$ that leads to the recursion relation for the free energy per site,

$$f(J) = g(J) + b^{-d} f(J') \tag{6.13}$$

with $N/N' = b^d$. The iteration has to be implemented together with that for the coupling, $J' = K(J)$. This equation can be written as the following recursion relation, $f(J_n) = g(J_n) + b^{-d}f(J_{n+1})$ that can be iterated as,

$$\begin{aligned} f(J_0) &= g(J_0) + b^{-d}f(J_1) \\ f(J_1) &= g(J_1) + b^{-d}f(J_2) \\ f(J_2) &= g(J_2) + b^{-d}f(J_3) \\ &\cdots \quad \cdots \\ f(J_n) &= g(J_n) + b^{-d}f(J_{n+1}). \end{aligned}$$

Substituting for $f(J_1)$, $f(J_2)$ and so on we get,

$$\begin{aligned} f(J_0) = {} & g(J_0) + b^{-d}g(J_1) \\ & + b^{-2d}[g(J_2) + b^{-d}[g(J_3) + b^{-d}[g(J_4) + \cdots]\cdots]] \end{aligned}$$

and

$$\begin{aligned} f(J_0) = {} & g(J_0) + b^{-d}g(J_1) + b^{-2d}g(J_2) \\ & + b^{-3d}g(J_3) + \cdots + b^{-nd}g(J_n) + b^{-nd}f(J_n), \end{aligned}$$

or

$$f(J_0) = \sum_{n=0}^{N} b^{-nd}g(J_n) + b^{-(N+1)d}f(J_{N+1}), \tag{6.14}$$

with $J_{n+1} = K(J_n)$. In equation (6.14), N is the number of iterations necessary to reach the attractor fixed point J^* of the corresponding phase, i.e., $J_{N+1} = J_N = J^*$. In general the last term in this equation can be neglected with respect to the sum. We can easily extend the calculation for the case where an external magnetic field is present. The result is

$$f(J_0, h_0) = \sum_{n=0}^{N} b^{-nd}g(J_n, h_n), \tag{6.15}$$

with $J_{n+1} = K(J_n, h_n)$, $h_{n+1} = H(J_n, h_n)$, $g(J_{n+1}, h_{n+1}) = Q(J_n, h_n)$. where the flow now is for the fixed point (J^*, h^*).

6.2.1 The specific heat of the 1d Heisenberg model

We now calculate the specific heat of the 1d Heisenberg model, in zero magnetic field, as an illustration of the method described above. For this purpose we consider the recursion relations, equations (6.11) obtained before. Since there is no finite temperature phase transition in this model for any non-zero temperature the iteration of the coupling is towards the high temperature, weak coupling attractor fixed point at $J^* = 0$. Then, starting from any arbitrary temperature $T_0 = 1/J_0$, we iterate this equation and obtain the values of the constant $G(J_n)$ at the image of the initial point, i.e., $G(J_0)$, $G(J_1) \cdots G(J^* = 0)$. Adding these values, conveniently multiplied by the scaling factors as in equation (6.15), we can calculate the free energy at any temperature $T_0 = 1/J_0$. The specific heat per site is obtained from $C = J^2(d^2J'/dJ^2)$ and is shown in figure 6.2. It has a maximum for $T/J \sim 2$ but no sharp feature since there is no phase transition.

6.2.2 The specific heat of the 3d Heisenberg model

The specific heat of the 3d Heisenberg model in the hierarchical lattice of figure 6.1 is shown in figure 6.3. Now there is a phase transition and the flows above and below T_c are to different attractors, but the same approach can be used. As can be seen in the figure, there is no divergence at the phase transition but a small cusp. However, this is consistent with the results obtained for this model. The correlation length exponent obtained at the unstable fixed point associated with the thermal phase transition is given by,

$$\nu = \ln b / \ln[(dJ'/dJ)_{J^*}].$$

Using equation (6.12) and the scaling factor $b = 2$, we obtain, $\nu = 1.4$. The exponent α that governs the singularity of the specific heat at the thermal phase transition is related to the correlation length exponent through the hyperscaling relation $2 - \alpha = \nu D_f$, where $D_f = 3$ is the fractal dimension of the hierarchical lattice. The value of ν given above implies a negative value for the exponent α consistent with the behavior shown in figure 6.3.

6.3 A simpler approach

The derivation above of the recursion relations for the Heisenberg model can be extended for the case of anisotropic interactions and disordered systems with

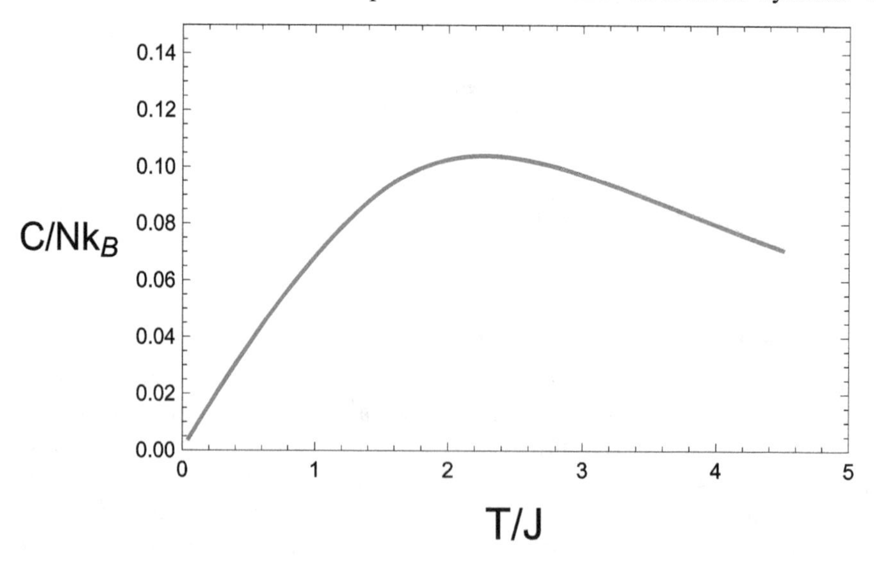

Figure 6.2. The specific heat/per site of the 1d Heisenberg model as a function of temperature.

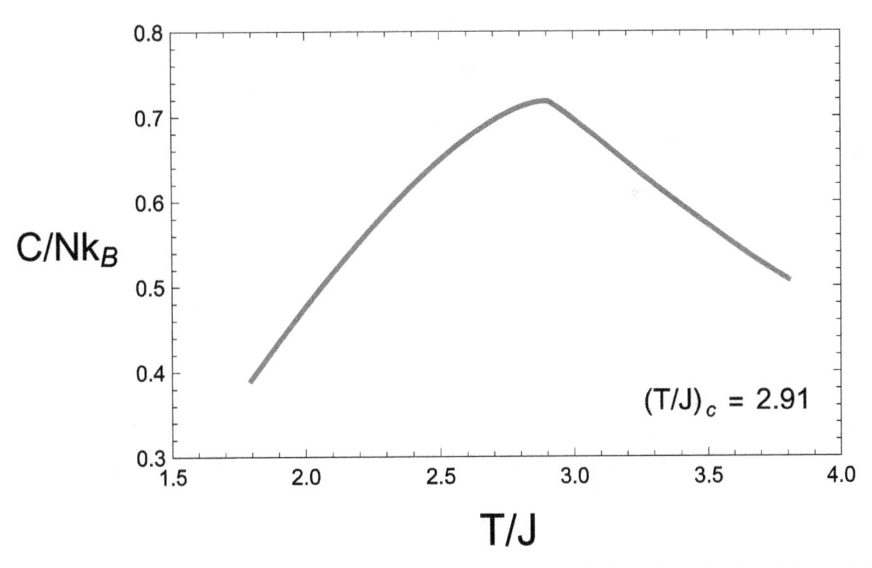

Figure 6.3. The specific heat/per site of the 3d Heisenberg model in the hierarchical lattice of figure 6.1 as a function of temperature, near the critical temperature.

different bonds. We now discuss a simpler, although more specific derivation due to Stinchcomb [3], based on the addition properties for spins-1/2.

For the cell with two spins, we have $\mathbf{s} = \boldsymbol{\sigma}_0 + \boldsymbol{\sigma}_2$. Taking the square of this relation, we get,

$$\boldsymbol{\sigma}_0 \cdot \boldsymbol{\sigma}_2 = \frac{1}{2}(\mathbf{s}^2 - 6), \tag{6.16}$$

where $\mathbf{s}^2 = 4s(s+1)$ and s can take the values 0 and 1 with multiplicity 1 and 3, respectively.

For the three-spins cell, we have $\mathbf{j} = \boldsymbol{\sigma}_0 + \boldsymbol{\sigma}_1 + \boldsymbol{\sigma}_2 = \mathbf{s} + \boldsymbol{\sigma}_1$. Taking the square, we get $\mathbf{j}^2 = 9 + 2(\boldsymbol{\sigma}_0 \cdot \boldsymbol{\sigma}_1 + \boldsymbol{\sigma}_1 \cdot \boldsymbol{\sigma}_2 + \boldsymbol{\sigma}_0 \cdot \boldsymbol{\sigma}_2)$, where $\mathbf{j}^2 = 4j(j+1)$ with $j = 1/2$ and $3/2$. Using equation (6.16) for $(\boldsymbol{\sigma}_0 \cdot \boldsymbol{\sigma}_2)$, we obtain,

$$(\boldsymbol{\sigma}_0 \cdot \boldsymbol{\sigma}_1 + \boldsymbol{\sigma}_1 \cdot \boldsymbol{\sigma}_2) = \frac{1}{2}(\mathbf{j}^2 - 9 - (\mathbf{s}^2 - 6)).$$

The states of the three-spins cell are classified in table 6.1.

The recursion relations are obtained from the RG procedure,

$$e^{G'+J'\boldsymbol{\sigma}_0 \cdot \boldsymbol{\sigma}_2} = \mathrm{Tr}_{\sigma_1} e^{2G+J(\boldsymbol{\sigma}_0 \cdot \boldsymbol{\sigma}_1 + \boldsymbol{\sigma}_1 \cdot \boldsymbol{\sigma}_2)}. \tag{6.17}$$

Given a value of s on the left hand side of the equations, we sum over all possible states j on the right hand site consistent with the values of s, to obtain,

$$(2s+1)e^{G'+J'(2s(s+1)-3)} = \sum_j (2j+1)e^{2G+J(2j(j+1)-2s(s+1)-3/2)}, \tag{6.18}$$

Table 6.1. The $2^3 = 8$ possible states of the three-spins cell. In the absence of a magnetic field the J_z states are degenerate, for a given j and s.

States of the three-spins cell		
s	j	j_z
0	1/2	1/2, −1/2
1	3/2	3/2, 1/2, −1/2, −3/2
1	1/2	1/2, −1/2

where the multiplicities have been taken into account. Specializing for the two values of s, $s = 0$ ($j = 1/2$) and $s = 1$ ($j = 1/2, 3/2$), with degeneracies $2j + 1$ we can obtain the two RG equations for J and G, equations (6.11), for the 1d Heisenberg ferromagnet.

6.4 Quantum phase transitions

The Heisenberg model studied in this chapter is a quantum model with thermal phase transitions at a critical temperature T_c for $d > 2$, but has no quantum phase transition [4]. For $d > 2$ it always has long-range order at zero temperature independent of the value of J. In general, to have a zero temperature phase transition the model needs to have at least two competing interactions, as in the Hubbard model we studied earlier (kinetic energy versus Coulomb repulsion). Since there is no entropy at $T = 0$, it is the competition between these interactions that can give rise to distinct ground states and a phase transition between them.

The simplest model exhibiting a quantum phase transition is the 1d Ising model in a transverse field [5],

$$H = -\sum_{ij} J_{ij} S_i^x S_j^x - h \sum_i S_i^z. \tag{6.19}$$

The nature of the competition is clear in this case. The ferromagnetic nearest neighbor exchange interactions ($J_{ij} = J > 0$) act to align the spins in the x-direction, while the Zeemann energy due to the transverse field is minimized when the spins point in the z-direction. In terms of the ratio h/J, when this is very large the total energy is minimized in the latter configuration. In the opposite case, i.e., for $h/J \ll 1$ the ground state is a long-range ordered ferromagnet with a finite magnetization $\langle S_x \rangle$ in the x-direction. As the ratio h/J increases, $\langle S_x \rangle$ vanishes at a quantum critical point (QCP) at $(h/J)_c$. This is a genuine quantum phase transition of second order since the order parameter $\langle S_x \rangle$ vanishes continuously as h/J approaches the critical value $(h/J)_c$. The transition is associated with a diverging length, the correlation length, that diverges at the zero temperature transition as

$$\xi = |g|^{-\nu}.$$

The quantity $g = (h/J) - (h/J)_c$ is the distance to the QCP and ν the correlation length exponent. Other critical exponents can be defined as in the case of thermal phase transitions. For example, the order parameter vanishes as

$$\langle S_x \rangle = |g|^{\beta},$$

which defines the critical exponent β. The order parameter susceptibility diverges at $T = 0$ with the critical exponent γ,

$$\chi_{xx} = |g|^{-\gamma}.$$

The magnetization $\langle S_x \rangle$ in the presence of an external field h_x in the x-direction behaves at the QCP $g = 0$ as,

$$\langle S_x \rangle = h_x^{1/\delta}.$$

The quantum critical exponents defined above, are not independent but obey scaling relations. As in the thermal case,

$$\begin{cases} \alpha + 2\beta + \gamma = 2 \\ \gamma = \beta(\delta - 1) \end{cases}$$

For quantum phase transitions, the thermal control parameter $g = (T - T_c)/T_c$ is replaced by $g = (h/J) - (h/J)_c$ however, in spite of many similarities quantum phase transitions have distinctive features. In the thermal case, the dimensionless free energy per unit volume is given by, $f = F/(k_B T)V$ where F is the free energy and V the volume. In this case the thermal energy $k_B T$ sets the energy scale for thermal fluctuations. In the quantum case, there are no thermal fluctuations and the energy scale has to be set by the interactions of the systems. In the case of the transverse Ising model this can be either the interaction J or the transverse field h. These quantities however renormalize under scale transformations. In particular at the QCP we can write,

$$J' = b^{-y} J. \tag{6.20}$$

Since the ratio $(h/J)_c$ is invariant under a scale transformation, the field h scales with the same exponent y at the QCP.

The exponent y can be identified with the dynamic exponent that governs the scaling of time in the renormalization process,

$$\delta t' = b^{z} \delta t.$$

Since in quantum problems fluctuations in energy and time are related by the Heisenberg uncertainty relation $\Delta E \Delta t \geqslant \hbar$ and since this is scale invariant, we arrive at the fundamental relation $y = z$. In quantum phase transitions static and dynamics are tied together by the Heisenberg inequality.

This has many important consequences. The dimensionless free energy per unit volume is now defined as, $f = F/JV$ and scales as [4],

$$f'(g) = b^{-(d+z)} f(g'),$$

under a length rescale by a factor b. Here $g = (h/J) - (h/J)_c$ and close to the unstable fixed point associated with the QCP, this scales as, $g' = b^{1/\nu}g$. Iterating these equations n times until $(b^n)^{1/\nu}g = 1$ and associating this length scale with the correlation length, i.e., $b^n = \xi$, we get,

$$f' = \xi^{-(d+z)}f(1).$$

Then the $T = 0$ free energy has a singular part given by,

$$f = |g|^{\nu(d+z)}. \tag{6.21}$$

On the other hand this singular part is used to define the critical exponent α, such that $f = |g|^{2-\alpha}$. Comparing this with the previous expression, equation (6.21), we get the *quantum hyperscaling relation* [4],

$$2 - \alpha = \nu(d + z). \tag{6.22}$$

When comparing this relation with its classical counterpart $2 - \alpha = \nu d$, we notice immediately that in the quantum case $d + z$ plays the role of an effective dimension. Effectively, the transition occurs in a hyperspace of $d + z$ dimensions, where to the space dimension d we add the time dimension z. This has dramatic consequences if we recall the existence in many phase transitions of an upper critical dimension d_c above which it is correctly described by a mean-field theory. Then in the case, not so uncommon, where $d + z > d_c$, the universality class of the quantum transition is associated with mean-field exponents.

6.4.1 The block method for the real space RG

Here we approach the 1d spin-1/2 transverse Ising model at zero temperature using the *block method* [6]. This consists in dividing a chain in blocks, for example, of two spins. The Hamiltonian, equation (6.19) is solved for the spins in the block and the two lowest energy states are identified as those of a spin-1/2 in an effective field h'. The new coupling between these effective spins is such that $J'S_{12}^{'x}S_{34}^{'x} = JS_2^xS_3^x$, where $S^x = \langle +|S^x|-\rangle = \zeta\langle +|S^{'x}|-\rangle$ and the states $|\pm\rangle$ correspond to the two lower energy states of a pair of spins (figure 6.4).

We shall not develop this method further since it has been well described and applied to several problems. We go directly to the RG equations that are obtained for the 1d transverse field Ising model. These equations can be written as,

$$\begin{cases} h' &= Jp[h/J] \\ J' &= Jq[h/J], \end{cases}$$

where $p[x]$ and $q[x]$ are functions that can be determined. This allows us to write the recursion relations in the form of a single equation for the ratio (h/J), we get,

$$(h/J)' = t[h/J], \tag{6.23}$$

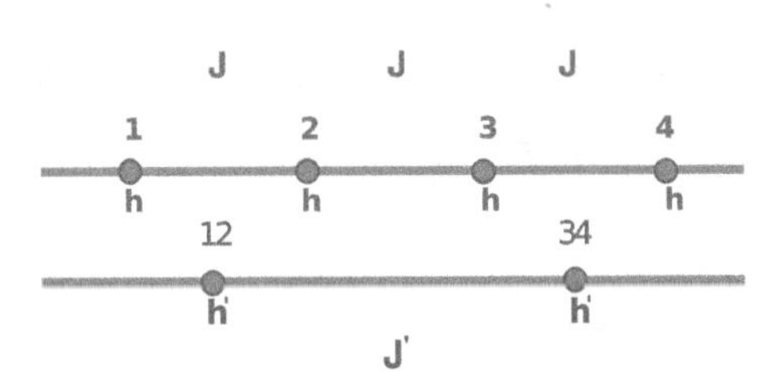

Figure 6.4. The block RG. The chain is divided into a block of two spins. The two lower states of this block are taken as the two possible states of a spin-1/2 in an effective field h'.

where $t[x] = p[x]/q[x]$. The equation $(h/J) = t[h/J]$, has two stable fixed points. One is at $(h/J) = 0$, and another at $(h/J) = \infty$ and they correspond to the attractors of the ordered ferromagnetic phase and that of the disordered phase, respectively. There is also an unstable fixed point at $(h/J) = (h/J)_c$ that is associated with the quantum phase transition from the ordered ($\langle S^x \rangle \neq 0$), to the disordered phase ($\langle S^x \rangle = 0$) at $T = 0$.

At this fixed point we have the following equation for the coupling,

$$J' = Jq[(h/J)_c] \equiv b^{-y}J, \tag{6.24}$$

and this allows us to identify the dynamic critical exponent,

$$z \equiv y = \frac{\ln q[(h/J)_c]}{\ln b}. \tag{6.25}$$

For the block of two spins with length scale factor $b = 2$, the numerical value obtained for the dynamic critical exponent, $z = 0.55$. This should be compared with the exact value $z = 1$. The 1d Ising model in transverse field at $T = 0$ maps in the $1 + 1$ Ising model for $T > 0$.

Let us consider the effect of temperature at the unstable fixed point of the model. Since J fixes the energy scale, temperature renormalizes at this fixed point as,

$$\left(\frac{T}{J}\right)' = b^z\left(\frac{T}{J}\right). \tag{6.26}$$

Notice that temperature scales away from the zero temperature fixed point and consequently it is a relevant variable in the RG sense. Indeed, temperature is necessarily relevant since there are no thermal phase transitions controlled by a quantum critical point. This also implies that in condensed matter there are no interactions that scale to strong coupling at the unstable quantum fixed point, i.e., the exponent $y > 0$ always. As we will see in the next chapter this is not necessarily the case in classical systems, especially in the presence of disorder.

The set of three equations close to the unstable fixed point,

$$\begin{cases} f'(g,(T/J)) &= b^{-(d+z)}f(g',(T/J)') \\ g' &= b^{1/\nu}g \\ \left(\frac{T}{J}\right)' &= b^{z}\left(\frac{T}{J}\right) \end{cases}$$

allow us to deduce the scaling properties of the free energy close to the quantum critical point [4],

$$f = |g|^{2-\alpha}\mathcal{F}\left(\frac{T}{|g|^{\nu z}}\right), \tag{6.27}$$

with $2 - \alpha = \nu(d + z)$ and $\mathcal{F}$ a scaling function. Notice the presence of the dynamic exponent z in the expression for the free energy. It is going to affect all thermodynamic quantities obtained as derivatives of the free energy. This is a unique feature of quantum phase transitions, since in classical transitions the dynamic exponent does not affect the thermodynamic properties. In the 2d Ising model in transverse field, the line $T_c = [(h/J) - (h/J)_c]^{\nu z}$ in the critical region of the phase diagram, i.e., for $(h/J) < (h/J)_c$ is a line of thermal phase transitions, as shown in figure 6.5. In the non-critical side of the phase diagram, the line $T^* = [(h/J) - (h/J)_c]^{\nu z}$ is a crossover line separating two regimes. There are no

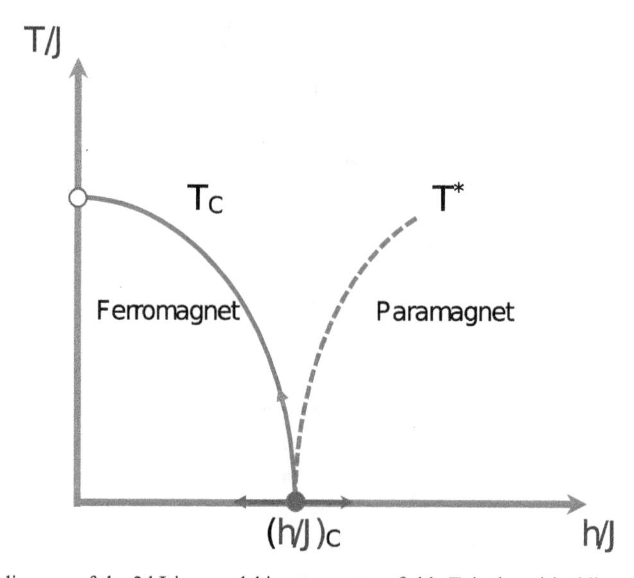

Figure 6.5. Phase diagram of the 2d Ising model in a transverse field. T_c is the critical line of phase transitions and T^* a crossover line. The fixed point at $T = 0$ (full circle) governs the quantum phase transition, i.e., the quantum critical exponents are calculated at this point. The fixed point at finite temperature (white circle) controls the thermal phase transitions along the critical line. The arrows indicate the flow of the RG equations.

singularities along this line only a change of behavior of the scaling functions. The system in the region $(h/J) > (h/J)_c$ is gapped and this gap vanishes at the QCP with the same exponent νz of the crossover line.

References

[1] Takano H and Suzuki M 1981 *J. Stat. Phys.* **26** 635

[2] Mariz A M *et al* 1985 *J. Phys. C: Solid State Phys.* **18** 4189

[3] Stinchcombe R B 1979 *J. Phys. C: Solid State Phys.* **12** 4533

[4] Continentino M A 2017 *Quantum Scaling in Many-Body Systems: An Approach to Quantum Phase Transitions* 2nd edn (Cambridge: Cambridge University Press)

[5] Pfeuty P 1970 *Ann. Phys.* **57** 79

[6] Jullien R 1981 *Can. J. Phys.* **59** 605

IOP Publishing

Key Methods and Concepts in Condensed Matter Physics
Green's functions and real space renormalization group
Mucio A Continentino

Chapter 7

Disordered systems

7.1 Introduction

In chapters 5 and 6, we have concentrated on systems without disorder, with the exception of the percolation problem that involves randomness in the position of the bonds. However, disorder is an important ingredient in many systems in nature. Also, it can be included on purpose to investigate the appearance of new universality classes. We have seen it breaks translation invariance requiring from the theoretical point of view to perform configuration averages of physical quantities to describe the behavior of the system. In many cases, this is not enough and one has to deal directly with the probability distributions and to distinguish between typical and average values. In this chapter we look at the effects of disorder using the real space renormalization group on hierarchical lattices. These are especially good approximations to real lattices in the presence of disorder. In many cases they give accurate predictions for the lower critical dimension of random systems, as we show below.

7.2 Random field models

The random field problem has raised much interest and has been extensively investigated since it can be realized in physical systems, as dilute antiferromagnets in an external field and disordered superconductors. The basic Hamiltonian is that of a Heisenberg or Ising ferromagnet with a random field,

$$\mathcal{H} = -\sum_{ij} J_{ij}\mathbf{S}_i \cdot \mathbf{S}_j - \sum_i \mathbf{h}_i \cdot \mathbf{S}_i. \tag{7.1}$$

The probability of random fields is such that

$$\begin{cases} \langle h_i^\alpha \rangle &= 0 \\ \left\langle h_i^\alpha h_j^\beta \right\rangle &= \langle h_i^2 \rangle \delta_{ij}\delta_{\alpha\beta}, \end{cases}$$

doi:10.1088/978-0-7503-3395-5ch7

where $\langle\cdots\rangle$ represents an average over all possible configurations of random fields. Details of the phase diagrams may depend on the specific form of the probability distribution. We focus here on the random field Ising model (RFIM) with a Gaussian distribution, $P(h_i) = (1/\sqrt{2\pi}h)\exp(-h_i/2h^2)$ centered at $\langle h_i \rangle = 0$ and with width $h = \sqrt{\langle h_i^2 \rangle}$.

The fate of the ferromagnetic state in the presence of the random field is determined by the balance of two energies. The Zeemann energy E_Z of droplets of size L consisting of spins pointing in the random field directions and the interfacial energy E_J arising from the exchange energy between droplets of ferromagnetically aligned spins [1]. The former varies as $E_Z \sim L^{d/2}$ where the factor (1/2) arises from the random nature of the field that points with the same probability in all directions. The interfacial energy is given by $E_J = L^{d-1}$ in the Ising case, since the domain walls are sharp and by $E_J = L^{d-2}$ in Heisenberg systems. As we have seen in the last chapter, these powers correspond to the scaling exponents of the magnetic field, with randomness taken into account, and of the exchange interactions close to the strong coupling attractive fixed point, i.e.,

$$\begin{cases} h' &= b^{d/2}h \\ J' &= b^{d-1}J \end{cases}$$

in the Ising case. As the droplets grow in size, there is a lower critical dimension $d_c > 2$ for Ising and $d_c > 4$ for Heisenberg systems above which, for small random field, the system will be ordered [1]. Above this critical dimension, the flow of the ratio (h/J) at $T = 0$, given by $(h/J)' = b^{(1-d/2)}(h/J)$, is towards the strong coupling attractor of the ferromagnetic Ising phase, as shown in figure 7.1.

Figure 7.1 shows a schematic phase diagram for RFIM in dimensions $2 < d < 6$, respectively, the lower and upper critical dimensions of this problem. As can be seen in the figure, close to the pure Ising fixed point at $(T/J)_c$ the random field is a

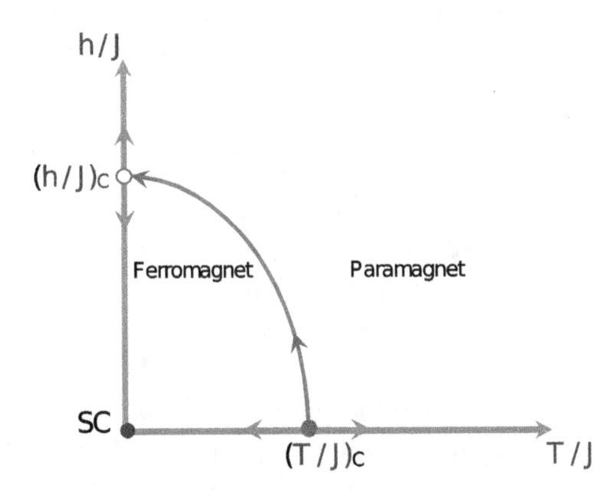

Figure 7.1. Schematic phase diagram of the random field Ising model for $2 < d < 6$. The random field is a relevant perturbation at the pure Ising fixed point $(T/J)_c$ (in red) and the ferro-paramagnetic transition is continuous down to $T = 0$. The thermal phase transitions along the critical line are governed by the RFIM zero temperature fixed point at $(h/J)_c$ (hollow) since temperature is irrelevant close to this fixed point. SC is the strong coupling attractor of the ferromagnetic phase.

relevant perturbation, since the flow of the RG equation is away from this fixed point [2]. There is a new universality class associated with thermal phase transitions along the critical line in the presence of the random field. This is governed by a special type of fixed point. It is located at $T = 0$, as in the case of quantum transitions. However, here we are dealing with a classical Ising system and the nature of this $T = 0$ fixed point is different from that of the quantum case. In the quantum case, temperature is always relevant and scales away from the $T = 0$ fixed point, such that thermal transitions are governed by a finite temperature fixed point. In the random field problem fluctuations due to disorder are stronger than thermal fluctuations and the thermal phase transitions in the presence of the random field are governed by the RFIM fixed point at $T = 0$, $(h/J) = (h/J)_c$. This arises since the width of the distributions of random fields renormalize to strong disorder at this fixed point, i.e., $h' = b^y h$, such that

$$(T/h)' = b^{-y}(T/h).$$

A real space renormalization group calculation on hierarchical cells with fractal dimension $D_f = 2$ and $D_f = 3$ has been implemented for the RFIM [3]. One starts from the equations for the transmissivities, equations (5.12) and (5.13) to obtain recursion relation for the fields. The cases of rectangular and Gaussian distributions of random fields with zero mean have been considered. In the former case the distribution of random fields evolves to a Gaussian characterized by the ratio between the fourth and second moment square, $\langle h_i^4 \rangle / h^4 = 3$. It is noteworthy that along the renormalization process the exchange couplings J_{ij} also become random. For $D_f = 3$, a phase diagram like that of figure 7.1 is obtained. The width of the distribution of random fields either decreases or increases along the renormalization process. In the former case the flow is to $\langle h/J \rangle \to 0$ and $\langle T/J \rangle = 0$ (the SC attractor in figure 7.1) that characterizes the ferromagnetic phase. In the latter, the flow is towards $\langle h/J \rangle \to \infty$ and $\langle T/J \rangle = \to\infty$, which is the attractor of the disordered paramagnetic phase. The critical (red) line in figure 7.1 separates these two distinct flows. The average value of $\langle h_i \rangle$ remains always fixed at zero during the renormalization process.

The real space RG calculation also allows us to obtain the critical exponents close to both the Ising and the RFIM fixed points. In the latter case, the exponent y is found to be $y \approx 1.48$. Close to the Ising fixed point we have $\phi/2 = \tilde{\nu}\tilde{y} \approx 1$. These calculations confirm the flow scheme in figure 7.1 and yield values for the critical exponents in agreement with other types of calculation. In the case $D_f = 2$, there is no ferromagnetic phase at finite temperatures consistent with the expected lower critical dimension $d_c = 2$

As for quantum transitions, a scaling theory can be developed near the fixed point at $T = 0$, $\langle h/J \rangle_c$ [2, 3]. The ground state energy per unit volume, $f = F/hV$ has at $T = 0$ the energy scale set by the width of the distribution of random fields, or the average of the exchange interactions. This also scales as $\langle J' \rangle = b^y \langle J \rangle$ since $\langle h/J \rangle$ is invariant at the fixed point. The scaling form of the free energy is given by,

$$\begin{cases} f'(g) &= b^{-(d-y)}f(g') \\ g' &= b^{1/\nu}g \end{cases}$$

where $g = |\langle h/J\rangle - \langle h/J\rangle_c|$. The exponent ν is the correlation length exponent at the random Ising fixed point (RIFP), $(T = 0, \langle h/J\rangle = \langle h/J\rangle_c)$. Defining a critical exponent α, by $f = |g|^{2-\alpha}$, we obtain as in the quantum case,

$$2 - \alpha = \nu(d - y). \tag{7.2}$$

As for quantum phase transitions, the hyperscaling relation is also modified. However, in the presence of the random field the effective dimension is *reduced* by the quenched disorder. The exponent y that renormalizes the coupling h and $\langle J\rangle$ in this classical disordered case has no relation to the dynamic critical exponent as in the case of quantum phase transitions. It can be obtained numerically in simulations and the results for three dimensions is very close to $y = 1.5$.

Since temperature is irrelevant and there is no crossover exponent at the RFIM fixed point $(T = 0, \langle h/J\rangle_c)$, it is interesting to consider the shape of the critical line close to this fixed point. Defining a $K = \langle h/J\rangle$ and $K_c = \langle h/J\rangle_c$, we can expand the recursion relations near the RIFP as,

$$\begin{cases} K_{n+1} &= K_c + b^x(K_n - K_c) + t_n^2 \\ t_{n+1} &= b^{-y}t_n, \end{cases}$$

where $t = T/\langle J\rangle$. The t^2 term in the first equation is the first analytic correction in temperature close to $T = 0$. Iterating these equations N-times, we obtain

$$K_N = K_c + \ell^x[K - K_c + at^2] - a(\ell^{-y}t)^2 \tag{7.3}$$

where $K_0 = K, t_0 = t, a = 1/(b^x - b^{-2y})$ and $\ell = b^N$ with b, as usual, the length scale factor. Notice that the *critical line*, $K - K_c + at^2 = 0$ is the set of points in the phase diagram that iterate to K_c after a large number of iterations, such that ℓ is very large. This critical line is analytic in temperature due to the irrelevance of temperature in the RG sense. It is very different from the shape of the critical line near the pure Ising fixed point, which is defined by a crossover exponent ϕ due to the relevance of the random field close to this fixed point.

For completeness let us consider an expansion of the RG equation close to the pure Ising fixed point.

$$\begin{cases} \mathcal{T}_{n+1} &= \mathcal{T}_c + b^{\tilde{x}}(\mathcal{T}_n - \mathcal{T}_c) \\ H_{n+1} &= b^{\tilde{y}}H_n, \end{cases}$$

where $\mathcal{T} = T/\langle J\rangle$, $\mathcal{T}_c = (T/\langle J\rangle)_c$ and $H = (h/\langle J\rangle)$. We use *tilde* exponents to make explicit that these refer to the finite temperature fixed point. Notice that the random field H (or h), as shown in figure 7.1, is a relevant perturbation at the Ising fixed point since $\tilde{y}>0$. We neglected the H^2 term in the first equation since this term gives a less singular contribution (see below). The solution of these recursion relations is obtained as before. It is given by,

$$\begin{aligned} \mathcal{T}_N &= \mathcal{T}_c + \ell^{\tilde{x}}[\mathcal{T} - \mathcal{T}_c] \\ H_N &= \ell^{\tilde{y}} H, \end{aligned} \tag{7.4}$$

where $\mathcal{T}_0 = \mathcal{T}$, $H_0 = H$ and $\ell = b^N$ as before. These equations imply the following expression for the free energy close to the Ising fixed point.

$$F/(Vk_BT) = |T - T_c|^{2-\tilde{\alpha}} \mathcal{F}\left[\frac{h}{|T - T_c|^{\tilde{\nu}\tilde{y}}}\right],$$

where $\tilde{\nu} = 1/\tilde{x}$ and $2 - \tilde{\alpha} = \tilde{\nu}d$. The shift exponent ψ that gives the shape of the critical line close to the Ising fixed point, $h = (T - T_c)^{\psi}$ is given by $\psi = \phi/2 = \tilde{\nu}\tilde{y}$, where ϕ is the crossover exponent. Since $1/(\tilde{\nu}\tilde{y}) < 2$, we can neglect the analytic contribution H^2 in the recursion relations for H that is sufficiently small.

7.3 Random spin chains

The chain of quantum spins described by the Heisenberg Hamiltonian with random nearest neighbors interactions has been subject of intensive study. One of the most successful approaches to this problem has been a perturbative real space renormalization group approach. The idea is originally due to Ma *et al* [4]. Consider a chain of quantum spins-1/2 interacting via nearest neighbor antiferromagnetic interactions as described by the Hamiltonian,

$$\mathcal{H} = \sum_{ij} J_{ij} \mathbf{S}_i \cdot \mathbf{S}_j, \tag{7.5}$$

where J_{ij} are random interactions with a probability distribution $P(J_{ij})$, with $J_{ij} > 0$. The method consists in finding the pair of spins with the strongest interaction, say $J_{23} = J$ as in figure 7.2 and remove this strongly coupled pair from the chain. This gives rise to an effective interaction (J'_{14}) between the neighbors of this pair, as shown in figure 7.2. The effective interaction is given by

$$J'_{14} = \frac{1}{2}\frac{J_{12}J_{34}}{J}. \tag{7.6}$$

Notice that the new effective interaction is always smaller than the one that has been removed. In this process, the distribution of interactions has changed and its cut-off has been reduced.

This procedure is then repeated until all interactions are removed, or in the case of an infinite chain, the probability distribution $P(J_{ij})$ attains a fixed point distribution given by [5]

$$P(J_{ij}) = \frac{\alpha}{J}\left(\frac{J}{J_{ij}}\right)^{1-\alpha}, \tag{7.7}$$

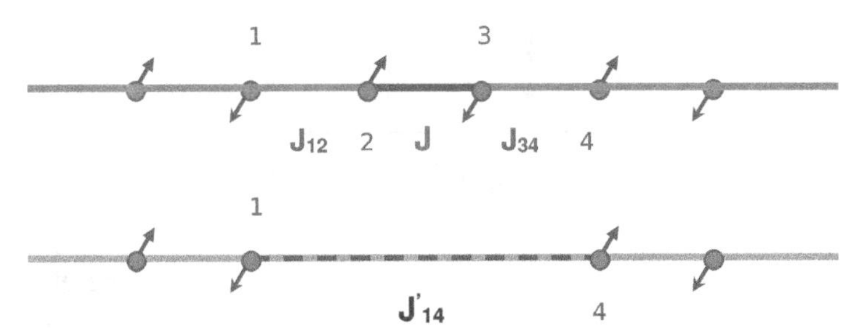

Figure 7.2. Random exchange antiferromagnetic chain with strongest coupling J. The pair coupled by this strong interaction is removed giving rise to an effective interaction J_{14}' between its neighbors (sites 1 and 4).

where

$$\alpha = \frac{-1}{\ln J}$$

varies with the cut-off J. This distribution characterizes a flow to what is known as an *infinite randomness fixed point* [5]. The magnetic phase of random Heisenberg antiferromagnetic chains associated with this distribution is known as a *random singlet phase*. It is a well-defined state with characteristic thermodynamic behavior. The susceptibility varies as

$$\chi(T) \propto 1/(T(\ln(J/T))^2). \tag{7.8}$$

The multiplicative nature of the recursion equation (7.6) suggests that the logarithmic of the interactions is more convenient to approach the problem of the random chain, since the distribution of interactions becomes very broad. The fraction of remaining spins $\rho = 1/L$ after the cut-off has been reduced to J is given by,

$$\rho = \frac{1}{L} = \frac{1}{|\ln(J/T)|^{1/\psi}} \tag{7.9}$$

for $J \ll T$ [5]. Instead of the usual scaling of energy, $J \sim 1/\tau \sim L^z$, we have a logarithmic relation between the length scale L of the chain of remaining spins and the actual cut-off J at this length scale. The exponent ψ is a new exponent that takes the value $\psi = 1/2$. The susceptibility is obtained assuming that the remaining spins have interactions much smaller than $k_B T$ and consequently behave as independent spins. This yields $\chi \sim \rho(1/T)$, and using equation (7.9) for ρ and $\psi = 1/2$, we get equation (7.8). The free energy is obtained, assuming equipartition, i.e., $F \sim \rho T$. The dominant contribution to the specific heat at low temperatures is $C = -T\partial^2 f/\partial T^2 = |\ln(J/T)|^{-(1+1/\psi}$.

It is possible to extend this approach to random exchange Heisenberg antiferromagnetic chains (REHACs) with arbitrary spins $S > 1/2$ [6]. The recursion relation in this case is given by,

$$J_{14} = \frac{2}{3}S(S+1)\frac{J_{12}J_{34}}{J}W_S(\beta J)$$

with $\beta = 1/(k_B T)$ and

$$W_S(x) = \frac{(2S+1)^2 - \sum_{j=0}^{2S}(2j+1)e^{-\frac{1}{2}j(j+1)x}\left(1 + \frac{1}{2}j(j+1)x\right)}{4S(S+1)\sum_{j=0}^{2S}(2j+1)e^{-\frac{1}{2}j(j+1)x}}$$

This recursion relation becomes problematic for $S > 1/2$ since the effective interaction generated by the removal of the strong coupled pair is not necessarily smaller than the removed interaction. Then, unless the initial distribution $P(J)$ has a very special form, the perturbative approach fails for $S > 1/2$.

In the case of REHACs of spins-1, for which [6]

$$J_{14} = \frac{4}{3}\frac{J_{12}J_{34}}{J},$$

this difficulty can be overcome. Instead of considering only pairs, it is also necessary to take into account *trios* of strongly coupled spins. Pairs or trios of spins are eliminated in such a way that the new effective interactions are always smaller than those that have been removed [7]. This guarantees the validity of the perturbative approach. A phase diagram of the spin-1 REHAC can be obtained starting from distribution of exchange interactions $P(J_{ij}) = (1/(1-G))\theta(1-J_{ij})\theta(J_{ij}-G)$, as shown in figure 7.3. These distributions have a gap that represents the amount of disorder in the system. As the gap G increases $(0 < G < 1)$ the system becomes less disordered. For $G = 0$ and only in this case, the initial distribution flows to the fixed point distribution given by equation (7.7). This shows the existence of random singlet phase in the spin-1 REHAC for strong disorder. For finite gaps, the phase diagram is shown in figure 7.4. For a gap smaller than $G_c = 0.45$ a Griffiths phase is obtained, and for $0.45 < G < 1$ there is a disordered Haldane phase. These phases

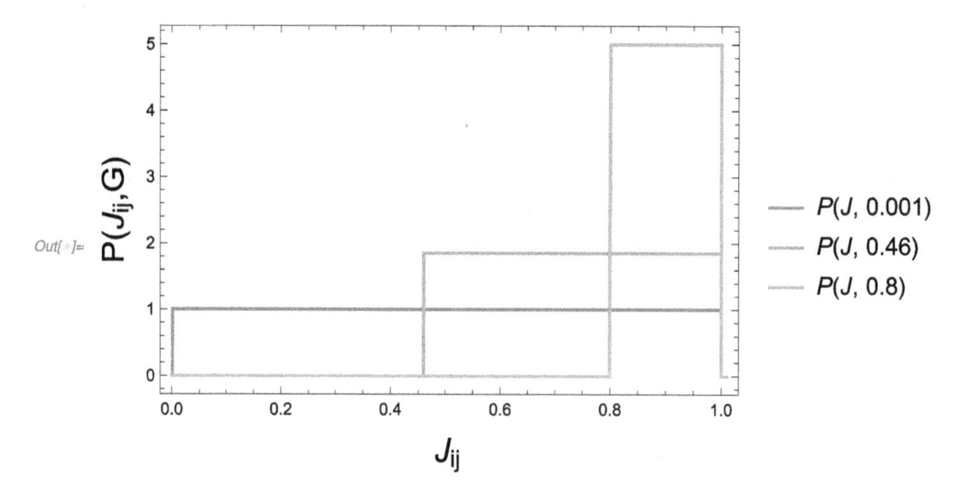

Figure 7.3. Distribution of exchange couplings $P(J_{ij}) = (1/(1-G))\theta(1-J_{ij})\theta(J_{ij}-G)$ used to obtain the phase diagram of a random spin chain with $S = 1$. The gap G is a measure of disorder that increases as G decreases. The random singlet exits only for zero gap.

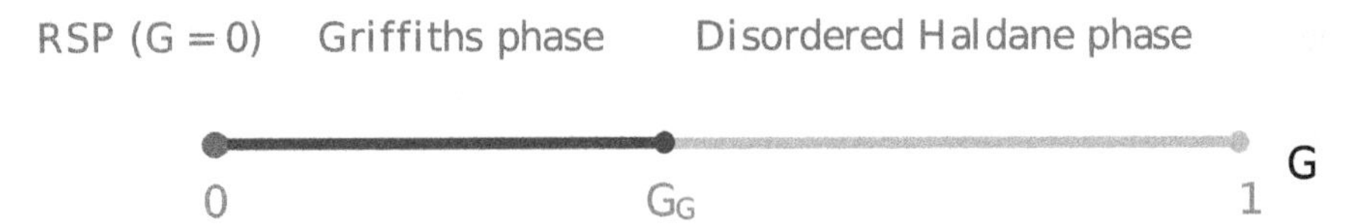

Figure 7.4. Phase diagram for the random antiferromagnetic spin-1 chain obtained from the distribution shown in figure 7.3. The random singlet phase exists only for gap $G = 0$. The Griffiths and disordered Haldane phases are described in the text.

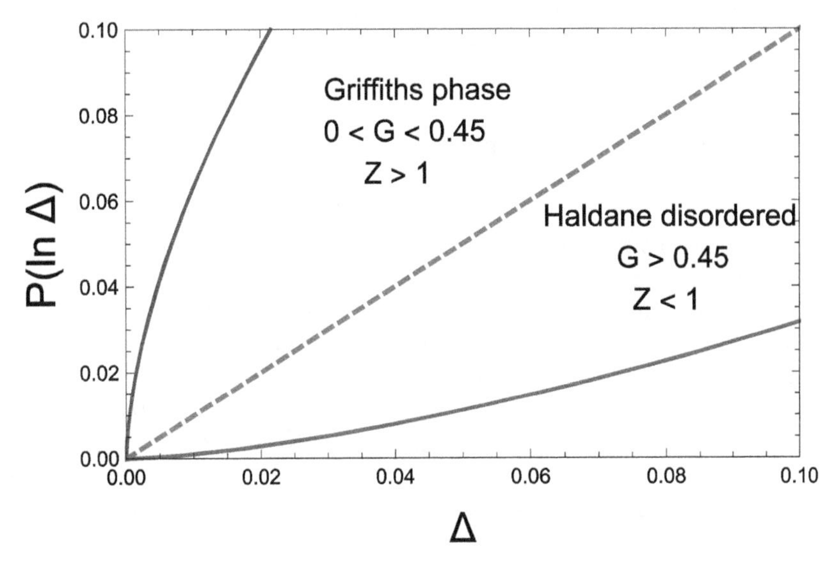

Figure 7.5. Phase diagram for the REHAC of spin-1. The line $Z = 1$, where $P(\ln \Delta) \propto \Delta$ separates a Griffiths from a disordered Haldane phase.

are better characterized considering the *first gap* distributions [7]. Starting with a large chain of size L and an initial distribution of interactions, as shown in figure 7.3, the perturbative renormalization group is carried on until a single pair of spins remains in the chain. The energy to promote this singlet pair to its first excited state is the *first gap* Δ. This procedure is repeated for several realizations of disorder and chains of different sizes L. A probability distribution of first gaps, or more conveniently of the logarithmic of first gaps can be obtained. For sufficiently large chains this distribution is independent of the size of the chain and given by $P(\ln \Delta) \propto \Delta^{1/Z}$, for small gaps. The *dynamic exponent* Z can take values from $Z \in (\infty, 0)$. The value $Z = \infty$ is attained at the quantum critical point (QCP) $G = 0$. As G increases, the dynamic exponent Z decreases. The system presents a Griffiths phase, with continuously varying dynamic exponents that depend on the distance to the QCP. The dynamic exponent reaches the value $Z = 1$ at a critical value $G = G_c = 0.45$ of the original distribution. For $G > G_c$, $Z < 1$ the system enters a disordered Haldane phase, as shown in figure 7.5. In this case there is *pseudogap* in the gap distribution. The susceptibility and specific heat have analytical behavior as temperature vanishes [7].

The problem of $S = 1$ random spin chains provides an elegant example of emergent symmetry, where the ground state has additional symmetries over those of the original Hamiltonian [8].

7.4 Anderson localization

The localization of electrons in random systems is an old problem but still with many unsolved questions. It has revealed along its development a surprising complexity which was in part anticipated since early studies. Even in non-interacting systems the prediction of the amount of disorder necessary to localize electrons, such that the system is an insulator at zero temperature is completely non-trivial and strongly dependent on its dimensionality. One of the reasons for this complexity lies in the configurational average of physical quantities in the presence of disorder. When disorder is sufficiently strong for localization to occur, it is not enough to average Green's functions or correlations functions that are related to physical quantities. These averages turn out to be meaningless due to the anomalous behavior of the probability distributions that in many cases have divergent moments. In order to have a meaningful description of the system, one requires a full knowledge of the probability distribution itself. Another way to tackle this problem, which we present here is to choose convenient quantities that have well behaved probability distributions, amenable to direct calculations [9]. This approach allows us to use the real space renormalization on hierarchical lattices without having to deal with the iteration of probability distributions, as shown by Shapiro [10].

Let us start with some dimensional analysis. The conductivity of a disordered metal is given by, $\sigma = ne^2\tau/m = ne^2\ell/(\hbar k_F)$ where n is the electron density and $\ell = v_F\tau$ is the free mean-free path. This has dimension $\sigma = (e^2/\hbar)L^{2-d}$. It is useful to define *dimensionless conductance* given by $g = \sigma L^{d-2}/(e^2/\hbar)$ and also a *dimensionless resistance* $\rho = 1/g$.

As shown by Anderson *et al* [9] in a disordered chain it is the quantity $\ln(1 + \rho)$ that has an additive mean and well-behaved statistical properties. Then for a wire with two disordered sections one finds,

$$\ln(1 + \rho) = \ln(1 + \rho_1) + \ln(1 + \rho_2), \tag{7.10}$$

where ρ_1 and ρ_2 are the dimensionless resistance of the sections of the wire.

Equation (7.10) replaces the usual sum for ohmic resistances in series. Generalizing for a wire consisting of b sections in series, all with the same resistance, we get

$$\rho' = (1 + \rho)^b - 1. \tag{7.11}$$

In an onion-like hierarchical lattice of fractal dimension D_f as in figure 6.1, we have to arrange $N = b^{D_f-1}$ of these wires in parallel with the same terminals. Since resistances in parallel add as their inverse, we have for the resistance of the hierarchical lattice,

$$\frac{1}{\rho'} = \frac{1}{\rho_1} + \frac{1}{\rho_2} + \cdots \frac{1}{\rho_N} = b^{D_f-1}\frac{1}{\rho}.$$

The final RG equation can be obtained in the form of a recursion relation,

$$\rho_{n+1} = b^{-(D_f-1)}[(1 + \rho_n)^b - 1]. \tag{7.12}$$

The fixed points of this equation are given by,

$$\rho^* = b^{-(D_f-1)}[(1 + \rho^*)^b - 1]$$

taking the length scale factor $b = 2$, we get:

- a stable fixed point at $\rho^* = \infty$ that is the attractor of the insulating phase;
- a fixed point at $\rho^* = 0$ that is stable for $D_f > 2$ and unstable for $D_f < 2$;
- for $D_f > 2$ an unstable fixed point at $\rho^* = 2(2^{D_f-2} - 1)$, at which occurs a metal–insulator transition.

Notice that $D_f = 2$ is the lower critical dimension for a metal–insulator transition. At $D_f = 2$ any disorder gives rise to an insulating state. These results actually coincide with those obtained in real systems and in a scaling theory of disorder-induced localization. An expansion of the RG equation close to the lower critical dimension, $D_f = 2 + \epsilon$, yields an unstable fixed point of $O(\epsilon)$ at $\rho^* = 2\epsilon \ln 2$. The correlation length exponent given by, $2^{1/\nu} = (d\rho'/d\rho)_{\rho=\rho^*}$ is obtained to first order in ϵ as, $\nu = 1/\epsilon$.

In this chapter we briefly reviewed some important problems in random systems. The usefulness of scaling ideas and of the real space RG has been emphasized.

References

[1] Imry Y and Ma S-k 1975 *Phys. Rev. Lett.* **35** 1399
Larkin A I 1970 *J. Exp. Theor. Phys.* **31** 784

[2] Bray A J and Moore M A 1985 *J. Phys. C: Solid State Phys.* **18** L927

[3] Boechat B and Continentino M A 1990 *J. Phys.: Condens. Matter* **2** 5277
Boechat B, dos Santos R R and Continentino M A 1994 *Phys. Rev.* B **49** 6404

[4] Ma S K, Dasgupta C and Hu C K 1979 *Phys. Rev. Lett.* **43** 1434
Ma S K and Dasgupta C 1980 *Phys. Rev.* B **22** 1305

[5] Fisher D S 1992 *Phys. Rev. Lett.* **69** 534
Fisher D S 1994 *Phys. Rev.* B **50** 3799
Fisher D S 1995 *Phys. Rev.* B **51** 6411

[6] Boechat B, Saguia A and Continentino M A 1996 *Solid State Commun.* **98** 411
Saguia A, Boechat B, Continentino M A and de Alcantara Bonfim O F 2001 *Phys. Rev.* B **63** 052414

[7] Saguia A, Boechat B and Continentino M A 2002 *Phys. Rev. Lett.* **89** 117202

[8] Quito V L, Hoyos J A and Miranda E 2015 *Phys. Rev. Lett.* **115** 167201

[9] Anderson P W, Thouless D J, Abrahams E and Fisher D S 1980 *Phys. Rev.* B **22** 3519

[10] Shapiro B 1982 *Phys. Rev. Lett.* **48** 823

IOP Publishing

Key Methods and Concepts in Condensed Matter Physics
Green's functions and real space renormalization group
Mucio A Continentino

Chapter 8

Topological systems

8.1 Introduction

In this chapter we study and review topological properties of one-dimensional condensed matter systems. Symmetry constraints give rise to non-trivial topological behavior with robust features. Topological non-triviality is accompanied by emergent properties and stands at the level of symmetry breaking. However, there are important differences. In the latter case the consequences are much more apparent since it introduces modifications that appear on the bulk. In the former, emergent properties occur mostly on the surface. Probably, for this reason, only recently were these features noticed and the study of topological systems, as topological insulators and superconductors started to receive the deserved attention. The theoretical description of the two phenomena also involve different approaches. In the case of symmetry breaking, this is generally accompanied by the appearance of an order parameter that in principle allows for a treatment within the Landau paradigm of phase transitions. The transition from a trivial to a non-trivial topological behavior in a system is much more subtle and is signaled by the abrupt change of a topological invariant.

8.2 One-dimensional Ising model in a transverse field revisited

In chapter 6, we introduced the one-dimensional transverse field Ising model (TFIM) as the simplest model exhibiting a quantum phase transition. Although we used this model to illustrate the Block RG method and obtain an approximate solution, it can be solved exactly [1]. Its solution corresponds to that of the classical Ising model in two dimensions since its dynamic quantum critical exponent $z = 1$ and consequently its effective dimension $d + z = 2$.

The solution of the model requires expressing the Hamiltonian of the 1d-TFIM, equation (6.19), in terms of fermion operators [2]. For this purpose we start rewriting the TFIM Hamiltonian in terms of raising σ^+ and lowering σ^- operators as,

doi:10.1088/978-0-7503-3395-5ch8

$$H = -J\sum_{i=1}^{N}(\sigma_i^+\sigma_{i+1}^+ + \sigma_i^-\sigma_{i+1}^- + \sigma_i^-\sigma_{i+1}^+ + \sigma_i^+\sigma_{i+1}^-) - h\sum_{i=1}^{N}\sigma_i^z. \tag{8.1}$$

The raising $\sigma^+ = (\sigma^x + i\sigma^y)/2$ and lowering $\sigma^- = (\sigma^x - i\sigma^y)/2$ operators recall spinless fermion operators since

$$\begin{cases} \sigma^+|\downarrow\rangle = |\uparrow\rangle \text{ and } \sigma^-|\downarrow\rangle = 0 \\ \sigma^+|\uparrow\rangle = 0 \text{ and } \sigma^-|\uparrow\rangle = |\downarrow\rangle, \end{cases} \tag{8.2}$$

in analogy with the fermionic case where

$$\begin{cases} c|1\rangle = |0\rangle \text{ and } c^+|1\rangle = 0, \\ c|0\rangle = 0 \text{ and } c^+|0\rangle = |1\rangle. \end{cases} \tag{8.3}$$

We are tempted to make the identification $\sigma^+ \to c$ and $\sigma^- \to c^+$, where a spin down in a site corresponds to a site with one fermion ($|\downarrow\rangle \to |1\rangle$) and a spin up to an empty site. However, while fermion operators in different sites anticommute, i.e., $\{c_i^+, c_j\} = \{c_i^+, c_j^+\} = \{c_i, c_j\} = 0$ if $i \neq j$, the spin operators *commute* i.e., $[\sigma_i^+, \sigma_j^-] = 0$ for $i \neq j$. Then, if we want to represent spin operators by fermions, it is necessary to circumvent this problem. This is accomplished by the Jordan–Wigner transformation, that multiplies the fermion operators by a non-local operator. The Jordan–Wigner transformations are given by,

$$\begin{cases} \sigma_n^+ = \left(\prod_{j=1}^{n-1}\sigma_j^z\right)c_n \\ \sigma_n^- = \left(\prod_{j=1}^{n-1}\sigma_j^z\right)c_n^+ \\ \sigma_n^z = 2c_n^+c_n - 1, \end{cases} \tag{8.4}$$

which finally allow us to write the TFIM Hamiltonian in terms of fermions operators. We get

$$H = -2h\sum_{n=1}^{N}c_n^+c_n - J\sum_{n=1}^{N}(c_n^+c_{n+1}^+ + c_n^+c_{n+1} - c_nc_{n+1}^+ - c_nc_{n+1}). \tag{8.5}$$

Equation (8.5) is a quadratic form that can be solved exactly using Fourier and Bogoliubov transformations. Alternatively, we can use the Greens function method. We define a Green's function $\langle\langle c_i; c_j^+\rangle\rangle_\omega$, that is the time Fourier transform of $\langle\langle c_i(t); c_j^+(0)\rangle\rangle$. The equation of motion for this Green's function is given by

$$\left\langle\left\langle c_i; c_j^+\right\rangle\right\rangle_\omega = \frac{1}{2\pi}\delta_{ij} + \left\langle\left\langle [c_i, H]; c_j^+\right\rangle\right\rangle_\omega.$$

Calculating the commutators, we get

$$\omega\left\langle\left\langle c_i; c_j^+\right\rangle\right\rangle_\omega = \frac{1}{2\pi}\delta_{ij} - 2h\left\langle\left\langle c_i; c_j^+\right\rangle\right\rangle_\omega - J\left\langle\left\langle (c_{i+1}^+ - c_{i-1}^+); c_j^+\right\rangle\right\rangle_\omega$$
$$- J\left\langle\left\langle (c_{i+1} + c_{i-1}); c_j^+\right\rangle\right\rangle_\omega.$$

Notice that a new Green's function has been generated by the equation of motion. This in turn obeys,

$$\omega\left\langle\left\langle c_i^+; c_j^+\right\rangle\right\rangle_\omega = 2h\left\langle\left\langle c_i^+; c_j^+\right\rangle\right\rangle_\omega + J\left\langle\left\langle (c_{i+1}^+ + c_{i-1}^+); c_j^+\right\rangle\right\rangle_\omega$$
$$+ J\left\langle\left\langle (c_{i+1} - c_{i-1}); c_j^+\right\rangle\right\rangle_\omega.$$

These equations can be Fourier transformed and we get,

$$\omega\langle\langle c_k; c_{k'}^+\rangle\rangle_\omega = \frac{1}{2\pi}\delta_{kk'} - 2h\langle\langle c_k; c_{k'}^+\rangle\rangle_\omega - J(e^{-ika} - e^{ika})\langle\langle c_{-k}^+; c_{k'}^+\rangle\rangle_\omega$$
$$- J(e^{-ika} + e^{ika})\langle\langle c_k; c_{k'}^+\rangle\rangle_\omega$$

and

$$\omega\langle\langle c_{-k}^+; c_{k'}^+\rangle\rangle_\omega = 2h\langle\langle c_{-k}^+; c_{k'}^+\rangle\rangle_\omega + J(e^{-ika} + e^{ika})\langle\langle c_{-k}^+; c_{k'}^+\rangle\rangle_\omega$$
$$+ J(e^{-ika} - e^{ika})\langle\langle c_k; c_{k'}^+\rangle\rangle_\omega,$$

which can be written as

$$(\omega + 2h + 2J\cos ka)\langle\langle c_k; c_{k'}^+\rangle\rangle_\omega = \frac{1}{2\pi}\delta_{kk'} + 2Ji\sin ka\langle\langle c_{-k}^+; c_{k'}^+\rangle\rangle_\omega \tag{8.6}$$

and

$$(\omega - 2h - 2J\cos ka)\langle\langle c_{-k}^+; c_{k'}^+\rangle\rangle_\omega = -2Ji\sin ka\langle\langle c_k; c_{k'}^+\rangle\rangle_\omega. \tag{8.7}$$

This system of two equations can be easily solve to yield

$$\langle\langle c_k; c_{k'}^+\rangle\rangle_\omega = \frac{1}{2\pi}\delta_{kk'}\frac{\omega - (2h + 2J\cos ka))}{\omega^2 - (2h + 2J\cos ka)^2 - 4J^2\sin^2 ka} \tag{8.8}$$

$$\langle\langle c_{-k}^+; c_{k'}^+\rangle\rangle_\omega = \frac{-2Ji\sin ka}{\omega^2 - (2h + 2J\cos ka)^2 - 4J^2\sin^2 ka}. \tag{8.9}$$

The energy of the excitations is given by the poles of the Green's functions,

$$\omega^2 - (2h + 2J\cos ka)^2 - 4J^2\sin^2 ka = 0,$$

or

$$\omega_{12} = \pm 2J\sqrt{(1 - (h/J))^2 + 2(h/J)(1 + \cos ka)}. \tag{8.10}$$

The calculation of the order parameter $\langle\sigma^x\rangle$ of the model is non-trivial, as can be noticed from equation (8.4). However, we can immediately obtain the correlation function $\langle c_k^+ c_k\rangle$ and from this the magnetization $\langle\sigma^z\rangle$, as can be easily verified from equations (8.4). For this purpose we require the Green's function $\langle\langle c_k;\, c_{k'}^+\rangle\rangle_\omega$ and use the leap theorem. From equation (8.4), we have

$$\langle\sigma^z\rangle = (1/N)\sum_n \sigma_n = (2/N)\sum_n \langle c_n^+ c_n\rangle - 1 = (2/N)\sum_k \langle c_k^+ c_k\rangle - 1. \tag{8.11}$$

In order to use the leap theorem, we rewrite equation (8.8) as,

$$\langle\langle c_k;\, c_{k'}^+\rangle\rangle_\omega = \frac{1}{2\pi}\frac{\omega - 2(h + J\cos ka))}{2\omega_1}\left(\frac{1}{\omega - \omega_1} - \frac{1}{\omega + \omega_1}\right), \tag{8.12}$$

where $\omega_1(k)$ is the positive root. We can then write,

$$\langle c_k^+ c_k\rangle = \mathcal{F}_\omega\{\langle\langle c_k;\, c_{k'}^+\rangle\rangle_\omega\}. \tag{8.13}$$

The magnetization is given by,

$$\langle\sigma^z\rangle = \frac{2}{N}\sum_k \int d\omega f(\omega)\frac{\omega - 2(h + J\cos ka)}{2\omega_1}[\delta(\omega - \omega_1) - \delta(\omega + \omega_1)] - 1, \tag{8.14}$$

where $f(\omega)$ is the Fermi function. Performing the frequency integral, we get,

$$\langle\sigma^z\rangle = \frac{2}{N}\sum_k\left[f(\omega_1)\frac{\omega_1 - 2(h + J\cos ka)}{2\omega_1} - f(-\omega_1)\frac{-\omega_1 - 2(h + J\cos ka)}{2\omega_1}\right] - 1. \tag{8.15}$$

This yields

$$\langle\sigma^z\rangle = \frac{1}{N}\sum_k \frac{1}{\omega_1}[\omega_1(f(\omega_1) + f(-\omega_1)) - 2(h + J\cos ka)(f(\omega_1) - f(-\omega_1))] - 1. \tag{8.16}$$

Using the expression of $f(\omega)$, we get

$$\langle\sigma^z\rangle = \frac{1}{N}\sum_k \frac{2(h + J\cos ka)}{\omega_1}\tanh(\beta\omega_1/2). \tag{8.17}$$

At $T = 0$ this is given by

$$\langle\sigma^z\rangle = \frac{1}{N}\sum_k \frac{h/J + \cos ka}{\sqrt{(1 - (h/J))^2 + 2(h/J)(1 + \cos ka)}}. \tag{8.18}$$

For $h = 0$, $\langle\sigma^z\rangle = 0$ and for $h/J = 1$, $\langle\sigma^z\rangle = 2/\pi$. The full zero temperature magnetization as a function of the transverse field is given by,

$$\langle\sigma^z\rangle = \frac{1}{2\pi}\int_{-\pi}^{\pi} dx \frac{h/J + \cos x}{\sqrt{(1 - (h/J))^2 + 2(h/J)(1 + \cos x)}} \tag{8.19}$$

and is shown in figure 8.1.

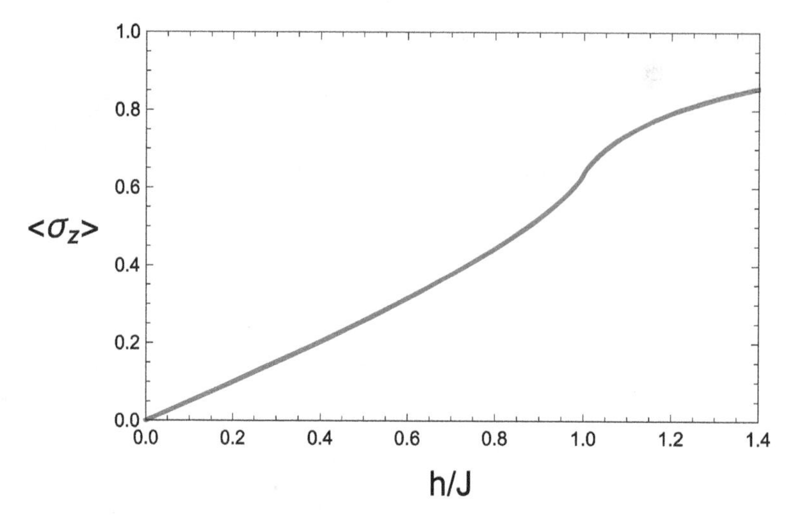

Figure 8.1. Zero temperature magnetization of the 1d transverse field Ising model, as a function of the transverse field. The quantum phase transition occurs for $(h/J)_c = 1$. Notice that this is not the order parameter spontaneous magnetization $\langle \sigma^x \rangle$ that vanishes at the quantum critical field.

We remark that the magnetization $\langle \sigma^z \rangle$ in the direction of the transverse field is **not** the order parameter of the system and for this reason it does not vanish at $(h/J)_c$. The order parameter is the magnetization $\langle \sigma^x \rangle$ and vanishes for $(h/J) \geqslant (h/J)_c$.

8.2.1 The quantum phase transition

The quantum phase transition of the $d = 1$ TFIM is associated with a soft mode in the spectrum of excitations at $k = \pm\pi$. Figure 8.2 shows the spectrum of excitations for $(h/J) = 0.45$ with the clear presence of a gap at $k = \pm\pi$, and at the critical value $(h/J)_c = 1$ where the gap vanishes and the phase transition occurs. Expanding the dispersion relation close to $k = \pm\pi$, we can rewrite it as,

$$\omega = 2J\sqrt{\left|\left(\frac{h}{J}\right) - \left(\frac{h}{J}\right)_c\right|^2 + \left(\frac{h}{J}\right)k^2},$$

where k stands for $k \to k \pm \pi$. Notice that at the quantum critical point $(h/J)_c = 1$, the dispersion relation, $\omega/2J = k$, such that the energy is linear in momentum and this defines the dynamical critical exponent $z = 1$ in agreement with what we pointed out earlier. Due to this linear dispersion, space and time scale with the same exponent and this guarantees the Lorentz invariance of the theory. The critical excitations are Dirac fermions. The effective dimension of the model is $d + z = 2$ and its quantum phase transition is in the same universality class of the classical two-dimensional Ising model. At $k = 0$ $(\pm\pi)$ the energy has a gap,

$$\Delta\omega = \left|\left(\frac{h}{J}\right) - \left(\frac{h}{J}\right)_c\right|^{\nu z}$$

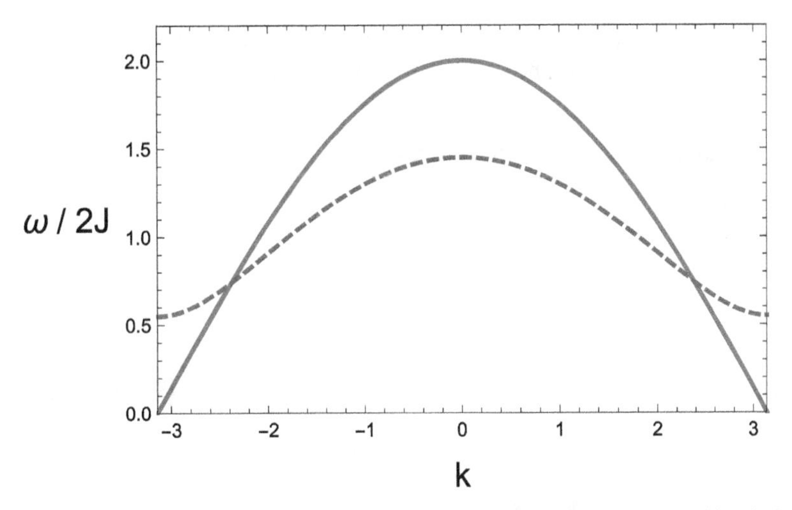

Figure 8.2. Dispersion relation of the fermionic excitations of the TFIM for $h/J = 0.45$ (blue dashed) and at the critical value $(h/J)_c = 1$. The gaps at $\pm\pi$ close at the critical value of h/J.

that vanishes at $(h/J)_c$ with the gap exponent $\nu z = 1$ that is the product of the dynamic exponent z and the correlation length exponent ν. Since $z = 1$, we find for the latter $\nu = 1$ in agreement with Onsager's result for the classical 2d Ising model.

The total, ground state energy of the 1d-TFIM, can be written close to the phase transition as,

$$\begin{aligned} E = \sum_k \omega_1(k) &= (JL/\pi)\int_{-\pi}^{\pi} dk\sqrt{(1-(h/J))^2 + 2(h/J)(1+\cos ka)} \\ &= (4JL/\pi)(-1+(h/J))\mathbf{E}\left[\frac{-4(h/J)}{(-1+(h/J))^2}\right], \end{aligned} \tag{8.20}$$

where L is the size of the large chain and $\mathbf{E}[x]$ is the complete elliptic integral. Incidentally notice that the magnetization, $\langle\sigma^z\rangle$, equation (8.19), can be obtained directly from the derivative of the total ground state energy with respect to the transverse field, i.e., $\langle\sigma^z\rangle = \partial E/\partial h$. As can be seen from figure 8.1, the magnetization as a function of the external field has an inflexion point at the quantum critical point $(h/J)_c = 1$. This is associated with a logarithmic singularity of the second derivative $\partial^2 E/\partial h^2$ of the singular part of the ground state energy with respect to the transverse field. This singularity arises since close to the QCP, the ground state energy, equation (8.20) behaves as $E \propto g^2 \ln|g|$, where $g = (h/J) - (h/J)_c$ is the distance to the QCP. This singular behavior corresponds to a logarithmic divergence of the specific heat of the classical 2d Ising model with which an exponent $\alpha = 0$ is associated.

We should warn the reader that in doing a Fourier transformation of the Green's function we had to assume periodic boundary conditions under certain assumptions. However, since we are interested in the limit of very large systems these become irrelevant, at least for the purpose of obtaining the spectrum of bulk excitations and thermodynamic quantities. In the next section we study properties of the TFIM for which the conditions at the ends of the chain become very important.

8.2.2 Topological properties of the TFIM

Let us consider a finite chain that supports a TFIM and has open boundary conditions. We start writing equation (8.5) as,

$$H = -2h \sum_{n=1}^{N} c_n^+ c_n - J \sum_{n=1}^{N-1} (c_n^+ - c_n)(c_{n+1}^+ + c_{n+1}) \tag{8.21}$$

We now define two new operators, both associated with a single fermionic state [3, 4]

$$\begin{cases} a_n &= c_n^+ + c_n \\ b_n &= i(c_n - c_n^+). \end{cases} \tag{8.22}$$

These operators represent fermions, as they obey the usual anticommutation rules for these particles, but they are also self-adjoints [4], i.e.,

$$\begin{cases} a_n^+ &= a_n \\ b_n^+ &= b_n. \end{cases}$$

They are known as *Majorana fermions*. In terms of these operators, the TFIM Hamiltonian can be written as,

$$H = iJ \sum_{n=1}^{N-1} b_n a_{n+1} + h \sum_{n=1}^{N} a_n b_n, \tag{8.23}$$

and an additional constant.

Let us examine this Hamiltonian in the limits of weak and strong coupling. As we saw in the previous chapter, the quantum phase transition in the 1d TFIM is governed by an unstable fixed point at $(h/J)_c = 1$. There are two attractors, the ferromagnetic strong coupling attractor at $h/J = 0$ and a weak coupling paramagnetic attractor dominated by the transverse field at $h/J = \infty$. We consider first the latter regime. In this case the transverse field h in equation (8.23) couples all Majoranas with their partners, at the same sites. The ground state can be pictured as in the top of figure 8.3. This phase is topologically trivial and can be described in terms of usual fermions.

In the ferromagnetic phase governed by the strong coupling fixed point, the interaction couples Majoranas in different sites, as shown in the lower panel of figure 8.3. We see from the Hamiltonian, equation (8.23) that two Majoranas, a_1 and b_N, remain uncoupled in this case. These two Majoranas can be used to define a single fermion state, $f = (1/2)(a_1 + ib_N)$ and $f^+ = (1/2)(b_N - ia_1)$ with occupation number $n_f = f^+ f$ that takes values 0 or 1. The total ground state energy is

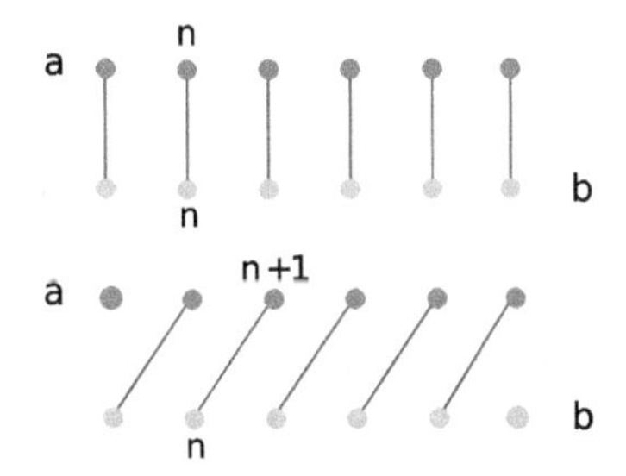

Figure 8.3. In the paramagnetic phase (top) the transverse field couples Majoranas at the same site. In the topological non-trivial ferromagnetic phase (bottom) that is governed by the strong coupling fixed point at $(h/J) = 0$, the interaction couples Majoranas at different sites leaving two uncoupled Majoranas, a_1, b_N, at the extremities of the chain [4].

independent of whether this state is occupied or not since it is does not contain a term $n_f = f^+f$ and consequently the system is degenerate. This fermion state is remarkable since it involves states on both extremities of the chain and consequently it is highly non-local. These properties make it useful as potential qubits for quantum computation.

The degeneracy of the *surface states* is due to the degeneracy associated with the Z_2 symmetry of the ferromagnetic Ising ground state. Notice that for $h/J > (h/J)_c$ the ground state is the spin polarized paramagnet. This state breaks the Z_2 symmetry and is non-degenerate. This paramagnetic state corresponds to the trivial topological state of the TFIM.

Then, in this model, the topological transition from a trivial to a non-trivial topological state occurs concomitantly with the symmetry breaking quantum paramagnetic-to-ferromagnetic phase transition. As we show below, this is not a necessary condition and a topological transition can occur without any symmetry-breaking. The topological transition is signaled by a topological invariant that changes values in the non-trivial and trivial topological phases. This is not an *order parameter* that vanishes continuously as in a second order phase transition. Since there is no latent heat associated they are not first order and do not fit in the usual Landau paradigm. However, we can still identify a diverging length accompanying this transition, so that it is a genuine continuous transition with which we can associate a set of critical exponents and characterize its universality class.

8.3 The Su–Schriefer–Heeger model

The Su–Schriefer–Heeger model (SSH) [5] is an important model for polyacetylene, an organic polymer represented by a linear chain where electrons can hop from site to site with alternating hopping terms, as in figure 8.4. The Hamiltonian of the model is given by,

$$\mathcal{H}_{ssh} = t_1 \sum_{i=1}^{N} a_i^+ b_i + t_2 \sum_{i=1}^{N-1} a_{i+1}^+ b_i + h.c. \tag{8.24}$$

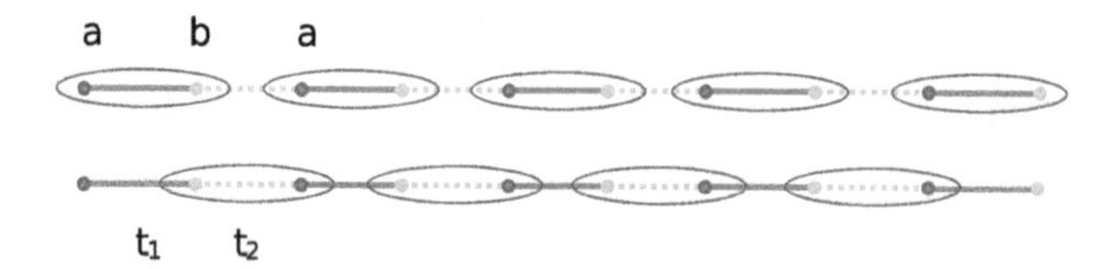

Figure 8.4. The two possible ground states of the SSH model. The terms t_1 and t_2 represent hopping terms and a and b the different sub-lattices. The top panel shows the trivial topological phase with $t_1 > t_2$ and topological invariant $W_1 = 0$. The lower panel shows the topological phase for $(t_2 > t_1)$ and that is characterized by the topological invariant $W_1 = 1$ (see text).

The operators a_i and b_i refer to electrons at site i of the sub-lattices a and b, respectively. In this equation $h.\ c$ means Hermitian conjugate terms.

The Green's function approach allows us to obtain the excitations of this model. For this purpose we introduce the sub-lattice Green's function $\langle\langle a_i; a_j^+\rangle\rangle$ and calculate its equation of motion,

$$\omega\langle\langle a_i; a_j^+\rangle\rangle = \frac{1}{2\pi}\delta_{ij} + \langle\langle [a_i, \mathcal{H}]; a_j^+\rangle\rangle.$$

The commutators are easily obtained and yield,

$$\omega\langle\langle a_i; a_j^+\rangle\rangle = \frac{1}{2\pi}\delta_{ij} + t_1\langle\langle b_i; a_j^+\rangle\rangle + t_2\langle\langle b_{i-1}; a_j^+\rangle\rangle.$$

Since a new Green's function has been generated, we also obtain

$$\omega\langle\langle b_i; a_j^+\rangle\rangle = t_1\langle\langle a_i; a_j^+\rangle\rangle + t_2\langle\langle a_{i+1}; a_j^+\rangle\rangle.$$

These equations in the case of the infinite system can be solved by Fourier transformation. Multiplying both sides by $e^{ikr_i}e^{-ik'r_j}$ and summing over (i, j), we get,

$$\begin{aligned}\omega\langle\langle a_k; a_{k'}^+\rangle\rangle &= \frac{1}{2\pi}\delta_{kk'} + (t_1 + t_2e^{ika})\langle\langle b_k; a_{k'}^+\rangle\rangle \\ \omega\langle\langle b_k; a_{k'}^+\rangle\rangle &= (t_1 + t_2e^{-ika})\langle\langle a_k; a_{k'}^+\rangle\rangle,\end{aligned} \tag{8.25}$$

where we assumed the hoppings t_1 and t_2 are real and a is the lattice spacing. These equations can be solved to yield,

$$\langle\langle a_k; a_{k'}^+\rangle\rangle = \frac{\delta_{kk'}}{2\pi}\frac{\omega}{\omega^2 - (t_1^2 + t_2^2 + 2t_1t_2\cos ka)} \tag{8.26}$$

and

$$\langle\langle b_k; a_{k'}^+\rangle\rangle = \frac{\delta_{kk'}}{2\pi}\frac{t_1 + t_2e^{-ika}}{\omega^2 - (t_1^2 + t_2^2 + 2t_1t_2\cos ka)}. \tag{8.27}$$

The poles of the Green's functions yield the excitations,

$$\omega = \pm\sqrt{t_1^2 + t_2^2 + 2t_1t_2 \cos ka}\,. \tag{8.28}$$

These dispersions are shown in figure 8.5. For $t_1 = t_2$ the system has a Dirac-like mode at $k = \pi/a$, otherwise, in the case of one electron per site, it is an insulator with a gap all along the Brillouin zone. The gap at $k = \pm\pi/a$ is given by

$$\omega(k = \pm\pi/a) = |t_1 - t_2|$$

and vanishes linearly when $t_1 \to t_2$. This defines the gap exponent $\nu z = 1$, where ν is the correlation length exponent and z the dynamic exponent. At the quantum critical point, $t_1 = t_2 = t$, and for $k \approx \pi$, the spectrum of excitations is given by,

$$\omega = tk,$$

where we replaced $k - \pi \to k$. This linear, Dirac-like spectrum implies a dynamic critical exponent $z = 1$ that together with $\nu z = 1$ leads to $\nu = 1$. At $t_1 = t_2$ there occurs a very special kind of phase transition. On both sides of the phase diagram $t_1 > t_2$ and $t_1 < t_2$ we find insulating states with the same symmetry. The quantum phase transition at $t_1 = t_2$ is a *topological transition.* The different insulating phases are characterized by distinct values of a topological invariant, as we show below. The non-trivial topological phase has surface modes, Majorana excitations, on both ends of a finite chain, as shown in figure 8.4. The topological trivial phase is a normal insulator with no surface modes, as shown also in figure 8.4. This can be verified explicitly writing the Hamiltonian in terms of Majorana operators as for the TFIM.

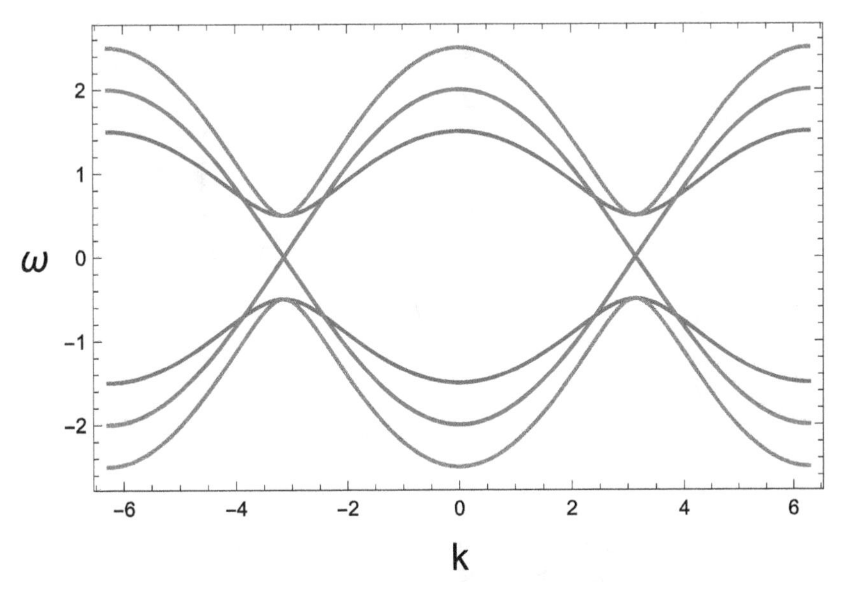

Figure 8.5. The dispersion relations for the SSH model, with $t_2 = 1$. In blue $t_1 = 0.5$; in gray $t_1 = 1.5$; in red at the topological transition $t_1 = t_2 = 1$.

The topological transition is outside the scope of the Landau theory of critical phenomena. There is no symmetry breaking and no order parameter. In spite of these idiosyncrasies, it is a genuine second order phase transition with critical exponents as the product νz that characterizes how the gap vanishes at the transition. The dynamic exponent z can be obtained from the spectrum of excitations at the transition that is Dirac-like and yields $z = 1$. The fundamental step is to identify the diverging length at the topological transition at $t_1 = t_2$ and obtain the correlation length critical exponent ν. In the p-wave superconducting chain model of Kitaev [6] and for the SSH model [7] this length has been identified as the *penetration length of the surface modes* in the non-trivial phase. The amplitude of the edge state wave function decays inside the chain as $\exp(-x/\xi)$ and ξ can be obtained numerically as the distance at which this decays to e^{-1} of its value at the surface. The length ξ depends on the distance to the topological transition, $\xi = |t_1 - t_2|^{-\nu}$, and diverges with the critical exponent ν, which is numerically determined to be $\nu = 1$, as shown in figure 8.6. This value is consistent with the result for the gap exponent $\nu z = 1$ and with the dynamical exponent $z = 1$. There are alternative ways for calculating ξ that yield similar results [8].

8.4 Bulk-boundary correspondence

Consider the ground state energy of the SSH model close to the topological transition,

$$E_{GS} = \sum_k \omega_k = \int dk\sqrt{|g|^2 + t^2k^2}, \tag{8.29}$$

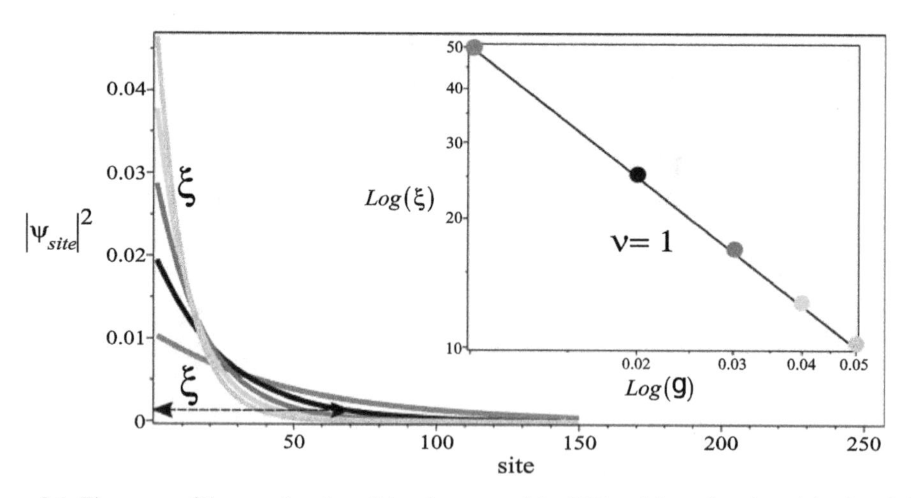

Figure 8.6. The square of the wave function of the edge states of the SSH model as a function of the sites. Solid lines are solutions for the a sublattice. The penetration depth ξ is shown for $g = |t_1 - t_2| = 0.02$ (black curve). At ξ the wave function satisfies the condition $\psi_n(\xi) = \psi_1/e$. The inset shows the penetration depth versus g. Different colors represent different values of g as depicted in the inset. The angular coefficient of the straight line is formally the critical exponent, $\nu = 1$. Reprinted from [7] with permission of Springer.

where $g = t_1 - t_2$. The most singular part of this integral goes as $E_{GS} \propto |g|^2 \ln|g|$ when $g \to 0$. This is associated with an exponent $\alpha = 0$. If we use the quantum hyperscaling relation $2 - \alpha = \nu(d + z)$ we verify this is satisfied with $\alpha = 0$ for the exponents $\nu = 1$, $z = 1$ and $d = 1$ of the SSH model. This quantum hyperscaling relation provides a *bulk-boundary correspondence* since it relates exponents of bulk quantities to that of the correlation length, which is a property of the edge states [7].

8.5 Spinor representation of the SSH model

For some cases it may be useful to introduce a Nambu or spinorial representation of the SSH model. Taking as a basis the vector (a_k, b_k), the matrix representation of the Hamiltonian of the SSH model is given by

$$\mathcal{H}_{ssh} = \sum_k (a_k^\dagger, b_k^\dagger)\mathcal{H}_{ssh}(a_k, b_k)^T, \tag{8.30}$$

or explicitly,

$$\mathcal{H}_{ssh} = \begin{pmatrix} 0 & t_1 + t_2 e^{ika} \\ t_1 + t_2 e^{-ika} & 0 \end{pmatrix} \tag{8.31}$$

This Hamiltonian has a sub-lattice or *chiral symmetry* since,

$$\sigma^{z-1}\mathcal{H}_{ssh}\sigma^z = -\mathcal{H}_{ssh} \tag{8.32}$$

where σ^z is the Pauli Matrix,

$$\sigma^z = \begin{pmatrix} 1 & 0 \\ 0 & -1 \end{pmatrix}. \tag{8.33}$$

We can also introduce a matrix representation for the Green's function as,

$$\mathbf{G}^{-1}(z, k) = z\mathbf{I} - \mathcal{H}_{ssh}(k). \tag{8.34}$$

Since $\mathbf{G}^{-1}(z = 0, k) = -\mathcal{H}_{ssh}(k)$, this has the same symmetry properties of the Hamiltonian and allows us to write topological invariants in terms of the Green's function.

8.6 Topological invariant for odd *d*-dimensional systems with chiral symmetry

The bulk invariant W_d for odd d-dimensional chiral systems was introduced by [9, 10], and [11]. W_d is given in terms of the Green's function by.

$$W_d = \frac{C_{d-1}}{2} \epsilon_{k_1 \ldots k_d} \int d^d k \; tr(\Gamma(G\partial_{k_1}G^{-1})\ldots(G\partial_{k_d}G^{-1})), \tag{8.35}$$

where $\epsilon_{k_1 \ldots k_d}$ is the Levi-Civita antisymmetric tensor,

$$C_d = -(2\pi i)^{-(d/2)-1}\frac{(d/2)!}{(d+1)!}.$$

Notice that W_d is evaluated at $\omega = 0$. The operator Γ that satisfies the chiral condition

$$G(\omega, k) = -\Gamma G(-\omega, k)\Gamma, \tag{8.36}$$

with $\Gamma^2 = 1$ is $\Gamma = \sigma^z$ as in equation (8.32) above. The integral in equation (8.35) is calculated at $\omega = 0$ and since $G(\omega = 0, k) = \mathcal{H}_{ssh}^{-1}$ we can write the invariant for the 1d SSH model as,

$$W_1 = \frac{-1}{4\pi i} \int_{-\pi/a}^{\pi/a} dk \; Tr\left(\sigma^z \mathcal{H}_{ssh}^{-1}(k) \partial_k \mathcal{H}_{ssh}(k)\right). \tag{8.37}$$

Calculating the matrices product in equation (8.37) and taking the trace, we get

$$W_1 = \frac{at_2}{4\pi} \int_{-\pi/a}^{\pi/a} dk \left(\frac{e^{ika}}{t_1 + t_2 e^{ika}} + \frac{e^{-ika}}{t_1 + t_2 e^{-ika}} \right). \tag{8.38}$$

In order to solve the integral above, we introduce the complex variable, $z = e^{ika}$ in terms of which it can be rewritten as,

$$W_1 = \frac{t_2}{4\pi i a} \left[\frac{1}{t_2} \int_C \frac{dz}{z + \frac{t1}{t2}} + \frac{1}{t_1} \int_C \frac{dz}{z\left(z + \frac{t2}{t1}\right)} \right], \tag{8.39}$$

where the contour of integration is a unit circle in the complex plane, as shown in figure 8.7. Rearranging, we obtain,

$$W_1 = \frac{1}{4\pi i a} \int_C dz \left(\frac{1}{z + \frac{t_1}{t2}} + \frac{1}{z} - \frac{1}{z + \frac{t2}{t1}} \right). \tag{8.40}$$

These integrals can be calculated using Cauchy's theorem. We find two cases:

- $(t_1/t_2) > 1$. In this case the pole of the first integral is outside the unit circle and the integral vanishes. The last two integrals are easy evaluated and we get,

$$W_1 = \frac{1}{4\pi i a}(0 + 2\pi i - 2\pi i) = 0.$$

- $(t_1/t_2) < 1$. Now the last integral vanishes since the integrand is analytic inside the contour. The first two integrals are obtained as before and we get

$$W_1 = \frac{1}{4\pi i a}(2\pi i + 2\pi i - 0) = 1,$$

where we took the lattice parameter $a = 1$.

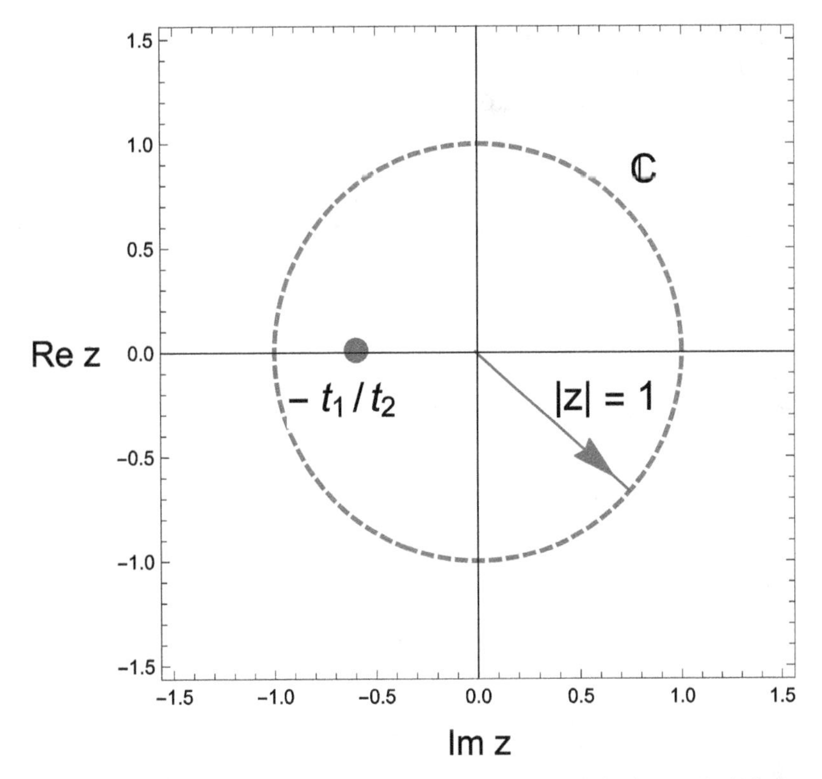

Figure 8.7. Contour of integration for equation (8.39) when the system is in the topological phase, i.e., for $|t_1/t_2| < 1$.

Then we have a quantity W_1, a *topological invariant* that can distinguish between the two phases of the SSH model. This quantity changes abruptly at the topological transition at $t_1 = t_2$. The phase with $(t_2/t_1) > 1$, with topological invariant $W_1 = 1$ is a topological insulator with surface modes, in this case Majorana excitations. Finally, the phase where $(t_2/t_1) < 1$, with topological invariant $W_1 = 1$ is a topologically trivial insulator with no surface modes. These phases are illustrated in figure 8.4.

References

[1] Pfeuty P 1970 *Ann. Phys.* **57** 79
[2] Kogut J 1979 *Rev. Mod. Phys.* **51** 659
[3] Dutta A, Aeppli G, Chakrabarti B K, Divakaran U, Rosenbaum T F and Sen D 2015 *Quantum Phase Transitions in Transverse Field Spin Models From Statistical Physics to Quantum Information* (Cambridge: Cambridge University Press)
[4] DeGottardi W, Thakurathi M, Vishveshwara S and Sen D 2013 *Phys. Rev.* B **88** 165111
[5] Shen S-Q 2012 *Topological Insulators: Dirac Equation in Condensed Matters* (Springer Series in Solid-State Sciences) (Berlin: Springer)
[6] Continentino M A, Caldas H, Nozadze D and Trivedi N 2014 *Phys. Lett.* A **378** 3340

[7] Continentino M A, Rufo S and Rufo G M 2020 *Strongly Coupled Field Theories for Condensed Matter and Quantum Information Theory* (Springer Proceedings in Physics vol 239) ed A Ferraz, K S Gupta, G W Semenoff and P Sodano (Cham: Springer) p 289
[8] Chen W, Rüegg A and Sigrist M 2017 *Phys. Rev.* B **95** 075116
[9] Gurarie V 2011 *Phys. Rev.* B **83** 085426
[10] Essin A M and Gurarie V 2011 *Phys. Rev.* B **84** 125132
[11] Volovik G E 2003 *The Universe in a Helium Droplet* (Oxford: Oxford University Press)

CPSIA information can be obtained
at www.ICGtesting.com
Printed in the USA
BVHW062003241021
619762BV00003B/89